Charles Seale-Hayne Library

University of Plymouth

(01752) 588 588

LibraryandITenquiries@plymouth.ac.uk

Zoophysiology   *Volume 23*

Coordinating Editor: D. S. Farner

Editors:
W. Burggren   S. Ishii   H. Langer
G. Neuweiler   D. J. Randall

Zoophysiology

Gerald L. Kooyman

Diverse Divers

Physiology and Behavior

With 87 Figures

Springer-Verlag
Berlin Heidelberg New York
London Paris Tokyo

Dr. Gerald L. Kooyman
University of California, San Diego
Scripps Institution of Oceanography
Physiological Research Laboratory
La Jolla, CA 92093, USA

Library of Congress Cataloging-in-Publication Data. Kooyman, Gerald L. Divierse divers: physiology and behavior/Gerald L. Kooyman. p. cm. – (Zoophysiology: v. 23) Bibliography: p. Includes index.

ISBN-13: 978-3-642-83604-6   e-ISBN-13: 978-3-642-83602-2
DOI: 10.1007/978-3-642-83602-2

1. Underwater physiology. 2.
Aquatic animals – Physiology. 3. Diving. Submarine – Physiological aspects. I. Title. II. Series. QP82.2.U45K66   1989   596'.01–dc19

© Springer-Verlag Berlin Heidelberg 1989

Softcover reprint of the hardcover 1st edition 1989

2131/3145-543210 – Printed on acid-free paper

To my mother and father, Virginia and Albert,
who let me keep rattlesnakes,
and to the late Kjell Johansen,
who proposed this book
and then encouraged me to carry on.

Preface

This book is not a conventional review of diving physiology. The coverage of the literature has been selective rather than encompassing, the emphasis has been on field studies rather than laboratory investigations, and the dive responses described are often discussed from the perspective of some of the flaws or weaknesses in the conclusions. Some of these points are of more historical interest to note how our concepts have evolved as we learn more about behavior and responses to natural diving in contrast to forced submersions in the laboratory. As a result there is a degree of evaluation of some experiments on my part that may seem obvious or controversial to the specialist. I have followed this plan at times in order to aid the reader, who I hope is often an untergraduate or graduate student, the nonspecialist, and the layman, in appreciating to some degree the level of dissatisfaction or skepticism about certain areas of research in diving physiology.

In view of historical boundaries in vertebrate biology, the subject is of broad enough importance to catch the interest of a wide audience of readers if I have done my job well. For example, of the major epochal transitions or events there have been in vertebrate history, three come immediately to mind: (1) The transition from aquatic to aerial respiration which ultimately led to a broad occupation of terrestrial habitats. (2) The development of endothermy. (3) Lastly and most germane to this book, the re-invasion of the sea over a broad time frame by diverse vertebrate groups. This radiation of vertebrates throughout the marine habitat is an inviting area to explore. The variety of constraints have in some instances resulted in common adaptation while in others they have diverged. That is the reason in part of the selection of some themes of discussion. One of these themes has been the way the O_2 store has been variable, while the rate of utilization appears to follow a common pattern of being low for most activities. Another problem common to all is how to deal with pressure through a range that is unimaginable to the terrestrial animal. Lastly, and the core of this book, is recognition of the diversity of behavior that has evolved in different groups of divers.

Because this is a book about the interrelationships of physiology and behavior of animals adpated to the marine habitat, a word should be said about sample sizes. There is no more challenging environment in which to work in the biosphere of the Earth. As a result, there are no conventional approaches in terms of equipment or sampling procedures. Inevitably sample sizes will be small. The investiagtor is dealing with large, often difficult-to-handle, and sometimes even rare species. All work on natural diving is done remotely, so that the animal cannot be observed directly. The experiments are time-consuming, yet constrained by seasonal patterns of behavior in which a very narrow window of time is available for the studies. Usually the work is done in isolated regions that require expensive logistics. But the most important and uncontrollable variable of all is the animal. To do studies of voluntary diving one needs a willing volunteer, and most are not. Often, if not usually, the animal, even in the most tractable species, may be erratic in its behavior; equipment in these harsh conditions fails all too frequently, or environmental conditions change from suitable to impossible. With patience and perseverance, it will all come together, but not often, When it does, that single animal and experiment are a treasure worth far more than any number of perhaps more statistically satisfying but unsound results. These frustrations can also be a delight because to obtain a full appreciation of the results an investigator must know the animal and its habitat well. That is the fun of it!

A word about the appendices. I have tried to be consistent with SI units, but much of the older literature and illustrations use other units. Where practical, I have converted the terms or used both. One of the most intractable is pressure, and Appendix I deals with the common conversions and equivalents. I have continued to use meters for depth rather than Pascals because of its ease of conversion between pressure and distances below the surface.

Appendix II lists all the species mentioned in the text, tables, and figures. Common names are used interchangeably with Latin binomials in the text but only the latter are used in tables, except Table 4.1.

Appendix III lists all the symbols used in the text.

Finally, there are too many people to whom I owe some debt of gratitude for all to be mentioned. However, a few that are exceptionally noteworthy should be named: first those that reviewed various chapters of the book. They are: Robert Blake, Michael Castellini, Paul Ponganis and Jesper Qvist. Others who helped through numerous discussions were the faculty and graduate students of the Physiological Research Laboratory, the northern elephant seal crew at UC, Santa Cruz, and Roger Gentry, who has been through some of the most difficult trials of instrument design, field testing, and manuscript preparation with me. In regard to manuscript preparation, B. Craig, P. Fisher, S. Henstock and P. Hooker made it all possible through their patient tolerance of handwritten drafts and revisions. Finally, and with sadness, I am indebted to the late Professor Donald S. Farner, who despite the serious illness that ultimately overwhelmed him, provided exceptional editorial advice. His polite and scholarly review and encouragement were greatly appreciated.

The writing of this book and some of the work mentioned in it were supported in part by USPHS HL17731 and NSF Division of Polar Programs grants DPP83-16963, DPP78-22999, and DPP86-13729, as well as NSF grant DCB84-07291 from Regulatory Biology, University of California, San Diego Academic Senate grant RJ148-S and the National Geographic Society.

LaJolla, Winter 1988/89   G. L. KOOYMAN

Contents

Chapter 1

A Perspective

Research on diving physiology of aquatic birds and mammals began as early as the mid-1800's, but an understanding of the subject did not blossom fully until the late 1930's and early 1940's as a result of the seminal works of L. Irving and P. Scholander. For the last 40 years there has been a great proliferation of studies of the response to breath-holding, ranging from experiments on great aquatic divers such as the elephant seal to the most unlikely candidates, such as the sloth. The subjects studied and the various experiments devised are a remarkable demonstration of the ingenuity of scientists to probe and tease information from the world around us. During these past 40 years of intense effort there has been one area of importance that has been explored desultorily, but not until recently has it moved, if not to center stage, at least out of the wings and into full view. The physiology and behavior of diving (not to be confused with forced submersions) is now coming into its own, and for some, is becoming the center of attention. Why has it taken so long? Probably because of the lack of technical means to explore this area, as well as tradition.

Traditionally, most physiologists have been trained in the laboratory to do laboratory problems. In the meantime, most field biologists were trained to study how animals behave using the most rudimentary tools of a notebook, binoculars, and careful observation. It seems to me that for many those boundaries are becoming less distinct now and "indoor" biologists are becoming more concerned about how the operations of the mechanisms that they have exposed function under natural conditions. Similarly, the "outdoor" biologist is becoming more interested in how the observed abilities of the subject are possible. That is what this book will be about – how marine tetrapods "work" under natural conditions.

More specifically, how the diving reptiles, birds, and mammals work in the functional sense of the word, as well as in the labor sense of the word. The central focus will be the physiology of diving with some final comments on general foraging behavior of these divers. The latter will be approached from a physiological point of view, rather than an ecological. Therefore, not much will be said about the divers impact on its prey, or its hunting strategy, but rather on how the anatomy and physiology of the animal sets the limits of its exploitation, and governs its behavior.

This approach is different from recently published reviews on diving, several of which have emphasized mechanisms obtained mostly from laboratory and forced submersion procedures (Butler and Jones 1982; Blix and Folkow 1983; Elsner and Gooden 1983). The only reviews of foraging behavior of aquatic vertebrates in which the behavior and physiology have been woven together has been the recent specific work on the Weddell seal (Kooyman 1981), and a more general behavioral study of fur seals (Gentry and Kooyman 1986). What I say in

Chapter 13 extends these studies to other vertebrate groups as well as to other species of seals.

In some ways the review may be premature because the study of both the behavior and physiology of diving, which is just budding, will probably come into full flower only in the next few years, thanks to the inventive application of microprocessors (Hill 1987). No doubt in this revolutionary era of electronics other equally useful tools will soon be available to the biologists who study behavioral and physiological adaptations under natural conditions.

1.1 Forced Submersion and Diving

At this early stage it is important to express a viewpoint that will be prevalent in this monograph, as well as to explain the selective use of certain terms. Experiments that involve the investigation of breath-holding by means of investigator-controlled stoppage of breathing by submersing the subject under water, placing a wet towel over the nose, tracheal clamping, drug-induced respiratory paralysis, or any other contrived means of initiating an episode of asphyxia will be termed forced, or controlled submersions, or breath-holds, but not dives. Such conditions have a few similarities to diving, and these similarities and differences will be discussed in later chapters. But the ground rules for this book are that they are not diving. This is not to say that some of the physiological responses observed are not similar to those that occur during diving, but rather that many may not be, so that the use of the expression has caused some confusion about the generality of these responses during diving.

The usage of the term diving will be restricted to voluntary submersions of animals that are of such a duration that there is a clear interruption in the ventilatory rhythm of the animal. Furthermore, it usually includes a departure from the surface to depth by active swimming efforts. Therefore, it *usually* incorporates two additional components besides just breath-holding – exercise and pressure. Both of these variables are not usually considered in laboratory studies. The reader no doubt has noted the "hedge" in this definition. It *usually* includes exercise, pressure, and a distinct break in the respiratory rhythm. There is a gray area of little-understood quiet submersions that will be correctly or incorrectly included as diving, but set apart by the specific statement that the act is a resting submersion.

Most, if not all species of marine mammals and reptiles are arrhythmic in their breathing; the elephant seal is the most extreme example of the mammals. At times its apneustic pauses while resting on land may be as long as 20 min. Such pauses occur while in the water also, and at these times the animals sink to the bottom in shallow water or drift some variable distance below the surface in deep water. Is this diving? The component of exercise is missing, but all other conditions are present. It is voluntary, increase in pressure is usually involved, and it is a departure from the respiratory rhythm of the awake animal. Until we know more about this type of breath-holding, or diving, it is considered as the latter and not just a breath-hold. This is not a serious problem at present because the condition has been studied so little that there will be only occasional reference to it in this monograph.

1.2 Content

The book is divided into two major topics, alluded to earlier. The first and largest section will be on the functional aspects of diving. It will begin with a selective review of breath-hold limitations as obtained from studies of tolerance to anoxia. This will be followed by a discussion of the effect of forced submersion on circulation. The experiments discussed will be those that in my view are most relevant at present to the understanding of freely diving animals, and/or those that I feel most competent on which to comment. These will be the criteria used throughout the book. There will be a further restriction to mammals, birds, and reptiles, and usually those that are marine. One reason for this is to avoid those species that rely to a considerable extent on cutaneous respiration (amphibians), and those that hibernate under water (some amphibians and reptiles). Both of those responses have resulted in adaptations that are probably not applicable to diving as previously defined.

An often neglected topic is that of effects of compression. Every diver experiences, on every submersion, some degree of compression. No doubt the variable of pressure has caused some striking anatomical and physiological adaptations to occur; the internal conditions of the diver are quite different from those of surface submersions or shallow dives. That is the reason that the subject is considered early rather than late as in most texts or reviews. It is hoped that by so doing the reader will continue to remain aware of pressure and how it might alter some of the results and responses, discussed elsewhere, that have been obtained only during surface submersions, or shallow dives.

After effects of pressure follows discussions of cardiovascular adaptations, blood chemistry, muscle chemistry and metabolism. This series of topics concludes the first of two phases in the book; the physiological responses to diving.

Part II, the labor of diving animals, will begin with a discussion of hydrodynamics (Chap. II) to set the stage for the general problems faced by a diver traveling through a dense and viscous medium compared to air. This will be followed by a discussion of the energetics of swimming. Compared with the topics in Part I, this is a relatively new area of research in which only few investigations have been done. The final topic will be diving behavior. This will focus on the specific characteristics of foraging dives.

1.2.1 Marine Divers

In addition to limiting this review to tetrapods, the focus will be mainly on marine divers. However, there are many works on freshwater divers that are important to this subject and some will be discussed.

3

For the purpose of an appreciation of the degree of diversity of marine divers from the various classes of vertebrates I will conclude this chapter with some brief comments about each class. Within the reptiles there have been four major invasions of the sea and one minor attempt at invasion. From early to late Mesozoic the plesiosaurs (pinniped like reptiles) and the ichthyosaurs (cetacean-like) were abundant. However, they failed, like the dinosaurs, to pass the Cretaceous/Tertiary boundary. Those reptiles that survived into the Tertiary were the marine turtles. Today there are eight species ranging in size from the 35–40 kg of the olive ridley, *Lepidochelys olivacea*, to the cosmopolitan leather-back, *Dermochelys coriacea*, in which the female mass may be as much as 500 kg and the males a ton.

Most successful among the present-day reptiles in terms of species diversification have been the sea snakes. There are more than 50 species; all are tropical and most are nearshore dwellers. The yellow-bellied sea snake, *Pelamis platurus*, is distinct in its pan-tropical distribution except for the Atlantic Ocean.

There are four species of saltwater crocodiles. These are also all tropical in distribution, and mainly coastal inhabitants.

The only lizard that has invaded the sea is the marine iguana, *Amblyrhynchus cristatus*. It is a nearshore herbivore which occurs only in the Galapagos archipelago. According to some (Dawson et al. 1977), it has no adaptations to an aquatic habit.

There are 17 families of sea birds, of which four families are divers. Best known, at least to the laymen, are the penguins, Spheniscidae, of which there are 17 species ranging in size from 1 kg of the little blue penguin, *Eupdyptula minor*, to 30 to 40 kg of the emperor penguin, *Aptenodytes forsteri*. Least known are the small diving petrels, of which there are four species within the family Pelecanoididae. The other marine divers are the murres and auks, Alcidae, with 22 species. Intermediate families that occur in both marine and freshwater habitats are the cormorants, Phalocrocoracidae; the loons, Gaviidae; the grebes, Colymbidae; and waterfowl, Anatidae. Most of these birds are found in temperate to polar waters.

There are three orders of mammals in which marine species have evolved. The only herbivores are the Sirenia, or sea cows. There are two families and four species within this group. All dwell in tropical rivers or coastal waters. However, the recently extinct Steller sea cow was a cold temperate species.

Within the Carnivora there are four families with aquatic species. The Mustelidae have two species, the well-known sea otter, *Enhydra lutra*, of temperate waters of Western North America and the gato marino, *Lutra felina*, of temperate waters of South America. The Phocidae contain 18 species of earless seals, all of which are polar to temperate except for the tropical monk seals. All 14 species of fur seals and sea lions, Otariidae, are temperate to cold temperate in distribution. Included are the two Galapagos species which latitudinally are tropical, but which usually occur at nearshore cold upwelling water. The one species of walrus in the family Odobenidae occurs only in cold temperate to arctic waters.

The Cetacea are the most diverse group of aquatic mammals, with nine families and 78 species. Most are pelagic, distribution of some is cosmopolitan, and the group is the most aquatically adapted of all mammals. Nevertheless, because of the technical difficulties of working with this group, both in captivity and in the wild, less will be said about their physiology and behavior than of the phocids and otariids.

Chapter 2

Tolerance to Anoxia

Before exploring the physiological responses to forced submersion and diving, I might ask the question of whether any kind of response is necessary, or whether, in addition to special physiological adaptations, it is likely that diving animals have an exceptional tolerance to anoxia that provides a safety margin if they were to extend their dives beyond the limit of their oxygen reserves? Reptiles are an ideal group in which to explore the adaptations to anoxia and/or diving because some members of this class do one or the other and perhaps both very well. Tolerance to anoxia in turtles was reviewed recently by Penney (1987), who concluded that their remarkable diving ability is a "spin-off" of their tolerance of hibernation. As you will see through the course of the following chapter, I do not think this is quite correct; many features of diving are quite different from those of torpor.

2.1 Reptiles

The tolerance of reptiles, especially turtles, to anoxia is notable. In a study of 70 species of reptiles in which the animals were ventilated on 100% N_2 until respiratory failure, it was found that turtles of all species and families exceeded the capacity for anoxia of any other reptiles (Belkin 1963) (Table 2.1). Of the seven families of turtles tested, the least resistant to anoxia were the sea turtles (Cheloniidae), which according to the experimental criteria reached their limit after 120 min of N_2 ventilation. Since some of the freshwater turtles hibernate under water during the winter time, it might be expected that they would have a greater anoxic capacity than the tropical sea turtles. However, it has been proposed that some green sea turtles, *Chelonia mydas*, may become torpid and rest on the sea floor for up to 3 months (Felger et al. 1976). Unfortunately, those

Table 2.1. Tolerance of anoxia in families of reptiles, based on time to respiratory arrest in ventilation with 100% N_2 at 22 °C. (Belkin 1963)

Class	No. of families	No. of species	Anoxic limit (range, h)
Turtles	7	27	1.9 to 33
Lizards	5	13	0.3 to 1.3
Snakes	4	29	0.4 to 2.0
Crocodiles	1	1	0.6

animals that do this may be a unique population of green sea turtles that has been decimated by Mexican fisherman; thus the physiological aspects of this problem may never be resolved.

The tolerance of the painted turtle, *Chrysemys picta*, to various states of asphyxia and anoxia shows broadly the capacity of some turtles for breath-holding, both while hibernating and while enduring forced submersions (Gatten 1981). After 2 days of forced submersion at 25 °C, the average whole body lactic acid (LA) concentration was 33 mMol l^{-1} (mM). After 2 weeks at 5 °C it again averaged 33 mM, and after a voluntary winter hibernation averaging 26 days at 0 to 8 °C the average whole body [LA] was 25 mM. One animal remained submerged for 67 days, at which time [LA] was 62 mM.

The most extended submersion of a sea turtle in which blood variables were obtained was a forced submersion of a green sea turtle (Berkson 1966). After 60 min the arterial O_2 was depleted, yet the turtle continued to breath-hold for another 5 h, even though it had been brought to the surface. Similarly, in hatchlings of loggerhead sea turtles, *Caretta caretta*, brain mitochondria can recover from at least 3.5 h of breathing 100% N_2 and return to oxidative activity (Lutz et al. 1980).

2.1.1 Whole Body Metabolism

These abilities of reptiles prompted a series of experiments over the years and in one of the early studies Belkin (1968) showed that there was a difference between stagnant anoxia and anoxic anoxia (Table 2.2), particularly in turtles. In the former, circulation was stopped by removal of the heart and the endpoint was respiratory arrest. In snakes and crocodiles, the difference in duration to endpoint was less than 1.5 times greater in anoxic anoxia than in stagnant anoxia. In turtles tolerance to stagnant anoxia was only twice as long as that of snakes and crocodiles, but it was 15 times shorter than some turtles' tolerance for anoxic anoxia. This experiment demonstrated the importance of metabolite transport to and from the organs.

The importance of anaerobic glycolysis was shown by experiments in which iodoacetic acid, the metabolic inhibitor of anaerobic glycolysis, was injected into

Table 2.2. Time (minutes, mean ± standard deviation) from first to last breath under conditions of acute anoxia at 22 °C, and 150- to 200-g animals. Numbers in parentheses are numbers of animals used. (Belkin 1968). Stagnant = cessation of blood flow; anoxic = continued blood flow, but no transport of O_2

Species	Stagnant anoxia	Anoxic anoxia
Caiman crocodilus	26.4±2.2(8)	33.0± 3.1(8)
Natrix fasciata	39.3±3.5(20)	62.2± 11.1(20)
Chrysemys picta	72.4±8.0(8)	1104 ±170 (8)
Pseudemys concinna	65.2±6.4(20)	900 ± 77 (20)

a musk turtle, *Sternothaerus minor*. Normally at 22 °C the turtle can tolerate anoxia for 12.2 h, but after the injection its limit was 0.32 h (Belkin 1968). Jackson (1968) duplicated this experiment on *Pseudemys scripta* in which 5 mg kg^{-1} of iodoacetic acid were injected prior to a submergence in 24 °C water. The turtle died after 60 min, whereas other turtles not injected were routinely submerged for 4–7 h.

Reduced metabolism due to the cold water is an important part of the ability of turtles to remain submerged for so long during winter dormancy. However, reduced metabolism is not so obvious a feature of submergence in warm water, and a series of calorimetric studies on turtles, *P. scripta*, held at 23.9 °C clearly showed an effect of O_2 availability on metabolism (Jackson and Schmidt-Nielsen 1966). When the turtles were ventilated on various concentrations, O_2 heat production was constant until the inspired O_2 concentration was below 5%. At an inspired O_2 of 3%, heat production was half that of controls; when breathing 100% N_2 it was 20% of controls.

If these turtles were submerged after breathing various concentrations of O_2 mixtures, a pattern similar to that in ventilation studies was observed. When the turtles were submerged after breathing 21% O_2, heat production steadily declined for 2 h, at which time it equaled that of turtles exposed to 100% N_2 before diving and turtles breathing 100% N_2. In later experiments Jackson (1968) showed that there was a close correlation between the decline in the fraction of lung O_2 and blood O_2 content and heat production (Fig. 2.1).

Respiration during submersion in water contributes a minor part, about 6% of total metabolism (Jackson and Schmidt-Nielsen 1966), while the turtle has sufficient O_2 available for aerobic metabolism. This contribution is also dependent upon the amount of dissolved O_2 (Fig. 2.2). When, however, the body stores of O_2 become exhausted and the metabolic rate (MR) has declined to 20% of rest-

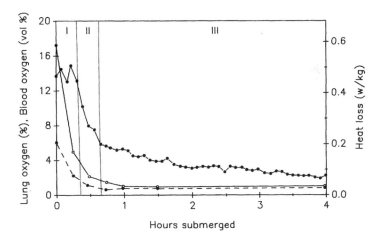

Fig. 2.1. Changes in O_2 content in blood and lung, and heat loss during submersion in turtles, *P. scripta*. Note that there is no change in heat loss for the first 20 min of submersion. *Solid line-closed circles* heat loss; *open circles* lung O_2; *broken line-closed circles* blood O_2 content. See text for stages I, II and III. (After Jackson 1968)

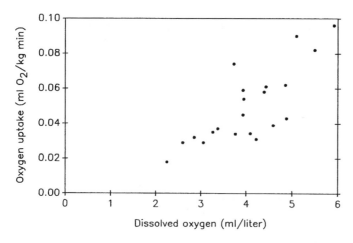

Fig. 2.2. Submerged turtle, *P. scripta*, uptake rate of dissolved oxygen in relation to water concentration of oxygen. Each *point* is uptake rate for 1 h. O_2 concentration is the mean for initial and final level during the hour. (Jackson and Schmidt-Nielsen 1966)

ing controls in air, aquatic respiration makes up to 30% of the total metabolic rate. For the earlier-mentioned marine turtles, which rest on the sea floor and are less tolerant to anoxic anoxia than other turtles, aquatic respiration may be an important factor in their ability for extended submersions, as it probably is in the yellow-bellied sea snake, *Pelamis platurus*. At 25 °C a 100-g snake will take up about 2 ml O_2 h^{-1} when O_2 gradient is 100 mmHg (13.3 kPa) (Graham 1974). However, their tolerance to low Pa_{O_2} is not exceptional, and the snakes succumb at a Pa_{O_2} <10–12 mmHg (Kooyman and Sinnett unpubl.).

A cause of the decline in MR (Fig. 2.1) was suggested to be due to the accumulation of LA (Jackson 1968), since increased [LA] may inhibit the conversion of pyruvate to LA by the law of mass action. The steady decline in MR through Phase III suggests a slow rise in [LA], which will set a limit to the duration of the submergence. The LA may become at least as high as 110 mM (Johlin and Moreland 1933), and the rate of increase is temperature dependent.

Jackson (1968) also showed that in Phases I and II (Fig. 2.1) the animals were capable of performing work (true diving) such as searching for food. However, by Phase III recorded muscle activity was depressed to the point that the turtles were nearly immobile and capable only of hiding or hibernating, and the occasional swim to the surface to breathe.

2.1.2 Cerebral Metabolism

In most vertebrates protection of the brain against an anoxic insult is paramount to that of all other organs. Turtles are remarkable in their ability to recover from extended periods of brain anoxia. Several ways in which they may achieve this have been proposed and discussed by Eugene Robin and his colleagues (Robin et al. 1979) specifically for the freshwater turtle *P. scripta*:

1. low brain metabolism,
2. large stores of adenosine triphosphate (ATP) and creatine phosphate (CP) in the brain,
3. utilization of non-glycolytic pathways,
4. enhanced glycolytic capacity, and
5. reduced energy requirements while submerged.

In a sense (5) comes full circle back to (1) but under the special condition of breath-holding. The previous results of whole body calorimetry had indicated a partial answer to question (5), that total energy consumption is greatly reduced during anoxia (Jackson and Schmidt-Nielsen 1966; Jackson 1968). Robin et al. (1979) attempted to answer some of the other five propositions by testing and comparing the metabolic results of rat and turtle brain slices.

They found that: (1) Low energy requirement was not likely because at 26 °C ATP production was equivalent in the two species. (2) Furthermore, inhibition of glycolysis by iodoacetate (Belkin 1963; Jackson 1968) showed that a store of CP and ATP was not likely and that glycolysis must be continuous. (3) Since the ratio of LA production and glucose consumption remained the same in the rat and turtle during aerobiosis and anaerobiosis it was presumed that nonglycolytic pathways were unlikely important contributors to brain energy production during anoxia. However, Robin et al. (1979) found that the rates of both aerobic and anaerobic glycolysis were higher in brain slices of the turtle than in those of the rat, indicating that the proposal (4), i.e., glycolysis capacity is enhanced, is likely. Furthermore, some of the augmented capacity was attributed to the high pyruvate kinase activity of turtle brain. Also, the lactate dehydrogenase (LDH) activity of turtle brain is double that of the rat (Miller and Hale 1968). Finally, since glycolysis accounts for only a fraction of the basal energy requirements of the brain, it is unlikely that the brain could survive long at this metabolic rate. Later evidence in the turtle, based on changes in [LA] in the brain in vivo, during ventilation with 100% N_2, indicate that ATP production declines substantially during anoxia (Lutz et al. 1984), so that, similarly to the reduced whole body metabolism mentioned earlier, there is also reduced brain metabolism.

2.2 Mammals

The emphasis in the previous discussion of turtles was on the ability of these animals to endure long-term anoxia. Mammals have no such tolerance, but in the short term, neonates and possibly divers have a higher level of tolerance than other mammals. In the following remarks I will discuss only divers, with a few comparisons to nondiving terrestrial mammals. Also, I will not discuss birds, for which there seem to be few investigations, and in which there is little tolerance to anoxia. Tolerance to anoxia in mammalian divers has been reviewed by Elsner and Gooden (1983), from whom I draw extensively.

Relative to all other organs the brain is the most intolerant to anoxia. This "weak link" will be the center of attention in the next few paragraphs. Other more tolerant organs such as the kidney and muscle will be discussed in Chapters 3, 7, and 8 about circulation in specific organs and metabolism while diving.

The degree of cerebral intolerance is fully appreciated from the results of experiments on arrested cerebral circulation. It was noted in these studies that loss of consciousness occurred in normal young men within 8 s (Rossen et al. 1943). In circulatory arrests exceeding 1 min the results are highly variable and dependent on a variety of conditions, but damage to the cerebral cortex may occur in even as short a time as 1 min; in the rat after 2 min of 100% N_2 ventilation electrocorticographic activity becomes isoelectric (Sick et al. 1982).

An intriguing variable influencing brain tolerance to anoxia is the concentration of LA. Myers (1977) discusses experiments on rhesus monkeys in which anoxic tolerance varied according to whether the animal was fasted or recently fed before the exposure. Fasted animals that are hypoglycemic tolerate the exposure better. Myers proposed that the reason was a difference in LA concentration in brain tissue. If brain tissue accumulate a LA concentration of 15–20 mM, then several hours later there is increased membrane permeability, tissue edema, and brain necrosis.

In contrast to brain, the muscle of dogs and cats can tolerate 15 h of occluded circulation; human muscle may be equally tolerant. For example, a common anesthetic procedure for "tennis elbow", with which many of the readers may be painfully familiar, is to eliminate blood flow to the arm during the 1 to 2 h required for surgical removal of the calcification on the tendon. This preserves the localization of the nerve-blocking agent in the surgical area. The procedure is called a Bier blook.

A different kind of exposure and one more pertinent to diving physiology is the time limit before dysfunction in the brain after the beginning of a breath-hold. A series of experiments on seals and dogs were performed in which the endpoint chosen was the onset of high voltage, slow wave electrical activity as recorded by an electroencephalogram (EEG). In the asphyxiated dog this endpoint occurs after about 4.5 min, which is close to the calculated limit of the body stores of O_2 if V_{O_2} persists at a resting level during asphyxia (Kerem and Elsner 1973 a). In yearling harbor seals of 25 to 30 kg, this endpoint is reached at 18.5 min after the beginning of forced submersion (Kerem and Elsner 1973), and in adult Weddell seals with arrested respiration, and immobilization by muscle paralysis using succinylcholine chloride, the endpoint was reached on average at 51 min.

The great difference between seals of different sizes, and between seals and dogs is attributed to the much larger O_2 stores and greater utilization thereof in the seal, as well as to striking alterations in distribution of blood flow. The latter is discussed in Chapter 3.

Greater utilization of the O_2 store and tolerance to low O_2 tensions in the seal were indicated from previously mentioned experiments as well. In the dog the EEG endpoint occurred at a arterial O_2 tension (Pa_{O_2}) of 14 torr (1.9 kPa) (Kerem and Elsner 1973), whereas in all seals, yearling harbor seals, and Weddell seal adults and pups, it occurred at about 10 torr (1.3 kPa) (Kerem et al. 1973;

Elsner et al. 1970). This compares to the human, where an EEG endpoint occurred at an internal jugular vein O_2 tension of 19 torr (2.5 kPa). It was noted in all studies that these endpoint blood tensions were consistent over a broad range of pH and P_{CO_2}.

In search of an explanation for this consistently greater tolerance in seals, Kerem and Elsner (1973 b) noted that the capillary density in the gray matter of the brain was 50% higher than in the mouse, cat, and man. The values were 515 cap mm^{-2} in the seal, and about 350 cap mm^{-2} in the other mammals. Presumably the higher density would reduce the diffusion distance within the tissue and therefore lessen the required O_2 gradient between capillary and tissue. However, another important variable would be cerebral blood flow (CBF). This was not measured, although a qualitative estimate using a catheter tip doppler flowmeter placed in the intravertebral extradural vein (see Chap. 3, Fig. 3.7, for diagram) at the base of the skull indicated that CBF was lower than the pre-dive level. This also is ambiguous because the pre-dive value may have been elevated well above rest in excited anticipation of the dive. Furthermore, a lower CBF would be opposite to the need of brain tissue in the face of a falling arterial O_2 tension.

2.3 Conclusions and Summary

For tetrapods, reptiles, especially freshwater turtles, have an exceptional tolerance to anoxia. It is interesting that some of the least tolerant are the sea turtles, which in general are tropical and are not exposed usually to conditions that would induce a winter torpor as in many temperate-zone turtles and tortoises. Two of the most important adaptations for torpor are a reduced metabolic rate and the anaerobic fermentation of glycogen.

Little is known about birds. Studies on mammals have been mainly on seals. The kidney of seals has an exceptional capacity for anoxia tolerance, which may be true of other organs as well. The brain of mammals is very intolerant to hypoxia, but the brain of seals continues to function normally in unusually low arterial tensions intolerable for other mammals.

Chapter 3

Forced Submersion and Circulation

Considering the ready accessibility of the domestic mallard, it is not surprising that it was the first subject of study of forced submersions. It is also a curious twist that the domestic mallard is derived not from a diving duck, but a dabbler, a species that feeds just below the surface in puddles, shallow ponds, or small streams. So much of our first information, as well as a good deal of later data on "diving" physiology, first came from a nondiver. Yet, for unknown evolutionary reasons this bird has remarkable tolerance to asphyxia, and has provided a considerable body of useful information about asphyxial reflexes. An extensive review of the contribution duck studies played in assessing asphyxial defense responses is presented by Andersen (1966), an investigator who himself made a significant impact on diving physiology by studying the domestic mallard. The review of Andersen has now been extended considerably by others, and Jones and Furilla (1987) in particular have presented a broader review on birds, but still with emphasis on the domestic mallard. In this work many of the ideas dealt with are discussed in this and later chapters.

3.1 Historical

The first studies were done by Paul Bert (1869), in which he compared the domestic chicken to the domestic mallard. He noted that the relative proportion of blood volume to body weight was greater in the mallard. After bleeding mallards to an equivalent blood volume of chickens, he found, not surprisingly, for reasons other than his conclusion of reduced O_2 store, that they succumbed after a breath-hold duration equivalent to that of chickens.

In the 1890's Richet (1894) and Bohr (1897, as cited in Andersen 1966), challenged this hypothesis on the grounds that the O_2 store above was insufficient to meet the requirements of these exceptionally long dives. What Richet and his collaborators noted was that the rate of O_2 disappearance from the air sacs decreased during the submersion, therefore, that V_{O_2} declined (Langlois and Richet 1898 a, b). Furthermore, immersion itself was important since birds suffocated in air after only 8 min of breath-hold compared to as much as 25 min if submersed (Richet 1899).

These observations led to a series of experiments by others shortly after the turn of the century, which investigated the reflex nature of the asphyxial response. Again, Andersen (1966) reviews these experiments and notes the suggestion by Bert that submersion bradycardia was under voluntary control.

However, Huxley (1913) found that bradycardia was elicited in decerebrate ducks as well. Several other studies followed dealing with the reflex characteristics of submersion asphyxia in mallards, on which Andersen (1966) comments. But it was not until the 1930's that Laurence Irving and Per Scholander, first independently, and then in the 1940's collaboratively, addressed the problem of asphyxia in true diving animals – penguins, seals, and porpoises. The brief review on the circulatory responses of diving animals to forced submersions that follows will begin with some of the results of the experiments that Scholander and Irving performed in the 1930's and 1940's.

3.2 Cardiovascular Responses to Forced Submersions

3.2.1 Heart Rate

The nearly universal response of bradycardia that accompanies forced submersions was noted at least as early as 1870 by Bert. This easily measured variable

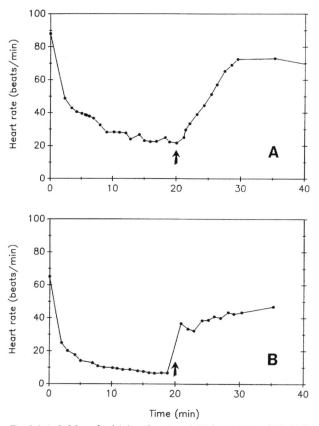

Fig. 3.1 A, B. Mean fetal (*A*) and maternal (*B*) heart rates of Weddell seals during forced submersions and recovery. Submersion begins at zero and ends at *arrow*. Sample size was 4. (After Liggins et al. 1980)

has since been used in almost all studies of forced submersion. The diversity of natural mammalian divers in which it has been observed, and the characteristic change with time during submersion was reviewed by Irving (1964). The rate and degree of bradycardia is highly variable from species to species. The most abrupt and intense decrease in rate is in the seal and the least, almost casual decline, is in the manatee. Even in the fetal seal bradycardia occurs during forced submersion of the adult (Fig. 3.1). The results are at slight variance with Elsner et al. (1970 a). Liggins et al. (1980) show both an initial higher rate, followed by a more rapid decline in rate, and then more rapid return to pre-submersion rate than does Elsner et al. (1970 a). The difference may be due to the means of restraint. In the work illustrated, the fully conscious animal was physically restrained on a tilt board during the submersions, whereas in Elsner et al. (1970) the seal was chemically restrained by the muscle-immobilizing agent succinylcholine chloride, which also produced paralysis of the respiratory muscles.

Marked changes in rate occur in the domestic mallard (Fig. 3.2), and in other birds such as penguins as well. For example, during forced submersions heart rates in gentoo and Adelie penguins declined from about 150 to 200 BPM to about 40 BPM within 1 min (Kooyman unpubl.). In reptiles the rate declines slowly, but the absolute value can be incredibly low. Berkson (1967) reported periods as long as 9 min in the green sea turtle when there was no heart beat. However, there were occasional pulsed increases of heart rate during struggling.

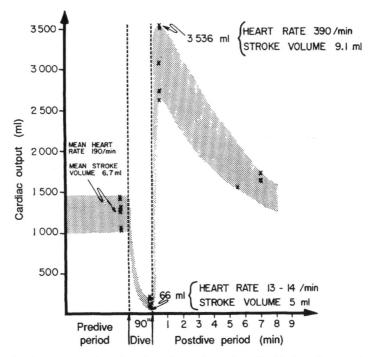

Fig. 3.2. Cardiac output, heart rate, and stroke volume in the domestic mallard before, during and after forced submersion. (Folkow et al. 1967)

17

3.2.2 Cardiac Output and Stroke Volume

The decline in heart rate indicates a substantial decline in cardiac output. How closely the degree of heart rate decline reflects the proportionate decrease in cardiac output is debatable. According to Elsner et al. (1964), Blix et al. (1976), and Murdaugh et al. (1961) the stroke volume remains constant in the sea lion, *Zalophus californianus*, gray seal, *Halichoerus grypus*, and harbor seal, *Phoca vitulina*. However, both Sinnett et al. (1978) and Zapol et al. (1979) showed that stroke volume declined by about 30 to 50% in the harbor seal and Weddell seal, *Leptonychotes weddellii*. The reason for these differences in results are obscure, but during forced submersions blood flow conditions are unstable. Some of the procedures used to measure cardiac output, which are generally designed for steady state flow, may be less reliable during submersion. Briefly, the sea lion study did not involve absolute measure but a relative one based on Doppler flow transmitter output of flow velocity characteristics which assume that vessel diameter remains constant. The method for the gray seal was not described, but was probably either by means of electromagnetic flowmeters or by the ratio of distribution of radionuclides. Both are relative measures. The early work on harbor seal was based on dye dilution; procedures and the results were highly variable from large SV increases to large SV decreases. The later study on harbor seals and the experiment with Weddell seals used a thermal dilution procedure which invariably gave decreased SV values.

Results from birds are equally confusing. Folkow et al. (1967) report little or no change in stroke volume of the domestic mallard, but the figure used in this review (Blix and Folkow 1983) shows a decrease in SV of 25% (Fig. 3.2). Addressing this question specifically, Jones and Holeton (1972) found, when using electromagnetic flowmeters, that SV *increased* during the dive from 15 to 35%.

These discrepancies may seem minor in regard to questions of heart contractility, or the degree in overall cardiac output change that occurs during forced submersions. The latter is well illustrated in Fig. 3.2, where it can be seen that CO drops about 90% during submersion. However, later discussions of voluntary dives and heart rate (Chap. 6) will show how much reliance is placed on this variable in estimating cardiac output. Such an estimate is valid only if there is a close relationship between heart rate and cardiac output.

3.2.3 Blood Pressure

Despite the considerable decline in cardiac output, the central arterial blood pressure is maintained in reptiles, birds, and mammals. However, in the green sea turtle after about 1.5 h of forced submersion, an exceptional period of time, the pressure was observed to decline to about 50% of the pre-dive level (Berkson 1966). Does this indicate reduced brain perfusion, and concomitant reduction in brain metabolism, as noted in freshwater turtles (Chap. 2)? An answer awaits further investigations.

In other vertebrates, the usual maintenance of blood pressure in the face of such a marked decline in cardiac output is due to an increase in peripheral vascular resistance (PVR). The marked level of increase was indicated by Butler and Jones (1982), who showed a concomitant fivefold rise in resistance to blood flow in the leg of the domestic duck with the decline in heart rate. Such an increase in PVR reflects a considerable reduction in peripheral blood flow.

3.2.4 Blood Flow

The resulting alterations in blood flow due to reduced cardiac output have been detailed for a few specific organs by means of Doppler and electromagnetic flow-meters, and more generally with radioactive microspheres. The flowmeters provide many values over a short or long period of time, and even in free-ranging animals. The disadvantages are that (1) placement of the flow cuffs requires major surgery, (2) only a few organs that have discrete arterial supply can be monitored, and (3) it assumes that vessel size remains constant because the primary measure is not volume of flow, but rate and volume must be extrapolated on a relative scale.

The microsphere procedures give relative information about flow to most, if not all organs, depending upon the injection site. The method is a count of the density of microspheres deposited in a given organ relative to a control value. Some disadvantages are: (1) only a few measurements can be made, (2) it measures the flow conditions at the time of injection so it provides discrete information but not the dynamics, (3) the results are dependent upon the size of microspheres, (4) usually the values are highly variable, no doubt due to the level of unsteady state oscillation in resting and submersed animals, and (5) the experiment is acute.

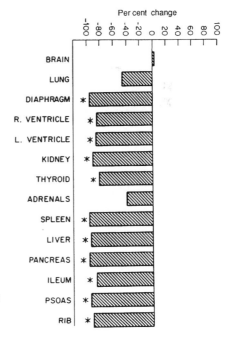

Fig. 3.3. Percent change in organ blood flow in Weddell seals during forced submersions. *Asterisks* denote probability that dive value differs from control (p<0.05). (After Zapol et al. 1979)

19

It can be seen that these measurements are technically difficult and expensive, so that few such investigations have been attempted. Only a few experiments have been done using the microsphere methods, and one using a radioactive salt solution. In general, these experiments have shown in both birds and mammals that for nearly all organs measured there is a substantial reduction in blood flow during forced submersions (Elsner et al. 1978; Johansen 1964; Jones et al. 1979; Zapol et al. 1979).

The most detailed study to date has been on the Weddell seal, in which it was found that cardiac output fell to 14% of pre-submersion values after 8 to 12 min of submersion and that there was a substantial reduction in blood flow to all organs except the brain (Zapol et al. 1979; Fig. 3.3). About 72% of the total cardiac output during the submersion was accounted for, and of that about 7.7% of the submersion cardiac output went to all parts of the brain compared to 1.1% in controls. Similarly, in the domestic mallard most organs except the heart have a substantial reduction in flow. Also, the brain has an increase in blood flow that becomes quite marked later in the dive (Jones et al. 1979).

3.2.5 Brain

Maintenance of blood flow to the brain of the Weddell seal was constant in all areas except the medulla, where there was a significant increase (Zapol et al. 1979). Observations from a smaller sample size in harbor seals showed that cortex and cerebellum blood flow were much less than the pre-submersion level at 2 and 5 min of submersion, but by 10 min cortex, cerebellum, and thalamus/hypothalamus were flows substantially higher (Fig. 3.4; Blix and

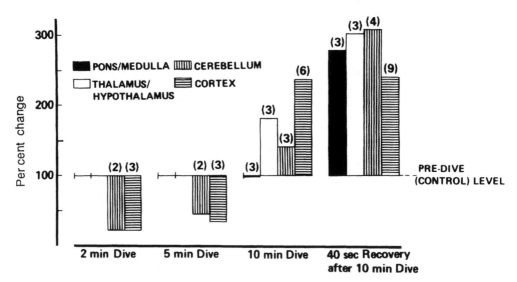

Fig. 3.4. Percent change in brain regional blood flow of the harbor seal during forced submersions. *Numbers in parentheses* are sample sizes. (Blix et al. 1983)

Folkow 1983), which contrasts with earlier experiments (Kerem and Elsner, 1973) discussed in Chapter 2, in which cerebral flow declined. An agreement with the Blix study was the increase in flow later in the submersion which was found in the duck, as mentioned above. It appears that some limit is reached where brain blood flow must be increased and presumably 12-min submersions were not long enough for this to occur in the Weddell seal, but 10- and 2.5-min submersions were sufficient in the harbor seal and domestic duck, respectively. In continuous measurements of internal carotid flow of the California sea lion, it was also shown that there is a steady rise in flow, much of it presumably to the brain, in the course of a submersion (Dormer et al. 1977).

3.2.6 Heart

Commensurate with the reduction in cardiac output and consequently work by the heart, coronary flow decreases proportionately (Blix et al. 1976; Zapol et al. 1979) (Fig. 3.3). A noteworthy exception is the mallard. No change in flow between the control and forced submersion was found in the mallard (Jones et al. 1979). At present, this seems inexplicable and a contradiction in the face of the reduced heart rate and cardiac output in domestic mallards when forced to submerge.

3.2.7 Kidney

During forced submersions in seals and ducks, renal blood flow virtually ceases, as shown by the microsphere method, Doppler flowmeters, and angiography. The cessation of blood flow, and the demonstration that it occurs at the level of the renal artery, have been dramatically illustrated directly by angiography in a submerged harbor seal (White et al. 1973), as well as by blood flowmeters in which the rapid rate of response is revealed by the onset of continuous, steady restriction of flow, or to release whenever the seal is submerged or surfaced (Elsner et al. 1966).

In the seal there is a close coupling between renal plasma flow and glomerular filtration rate (Fig 3.5). Thus, the restriction in blood flow causes a marked reduction in glomerular filtration rate and urine production. Consequently, renal excretion of sodium and urea decrease during forced submersions (Bradley et al. 1954; Schmidt-Nielsen et al. 1959).

3.2.8 Skeletal Muscle

The well-maintained abdominal aortic pressure in the face of a considerable drop in CO during forced submersions is largely due to reduction in blood flow to skeletal muscle. During forced submersions blood flow virtually ceases as shown in angiographs of the hindquarters of harbor seals (Bron et al. 1966) and from microsphere distribution in the limited measurements obtained (Fig. 3.3).

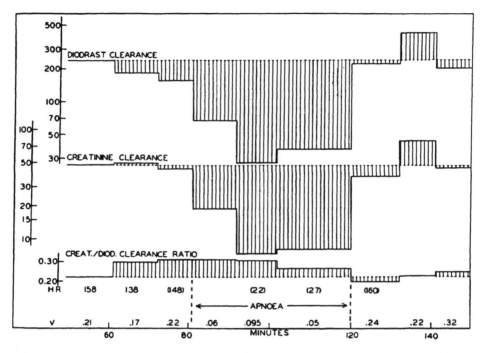

Fig. 3.5. Renal plasma flow (diodrast clearance, ml min^{-1}), glomerular filtration rate (creatinine clearance, ml min^{-1}), filtration fraction (creatinine/diodrast ratio), heart rate, and urine flow (V, ml) at rest and during asphyxia in the harbor seal. (Bradley and Bing, 1942)

The few muscle biopsies that have been obtained during dives best illustrate the functional character of restricted flow. The decline in muscle O_2 is mirrored in the increase in lactic acid (LA) (Fig. 3.6) in the harbor seal. Tightness of control in this species is further demonstrated by the marked response in blood LA concurrent with increase in LA and decrease in O_2 in muscle. While muscle O_2 declines rapidly and exponentially, arterial O_2 declines slowly and linearly (Fig. 3.6) over the same time period. Similarly, muscle LA increases very rapidly, while there is almost no change in arterial blood.

3.2.9 Uterus

In two separate studies using different procedures the results agree that uterine blood flow in the pregnant Weddell seal is maintained (Elsner et al. 1970 a; Liggins et al. 1980). The flowrate of uterine blood remains approximately the same during as it was before submersion. Consequently, the percent of total dose of microspheres increased sixfold over pre-submersion control doses during forced submersions (Liggins et al. 1980). These authors noted that despite a marked bradycardia in the fetus the mean arterial pressure remained high, presumably due to a broad vasoconstriction.

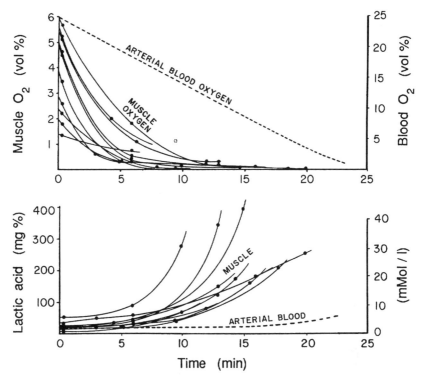

Fig. 3.6. Oxygen depletion and accumulation in muscle and blood in the harbor seal during forced submersions. (After Scholander et al. 1942)

3.2.10 Arteriovenous Anastomoses

In the course of a submersion the arteriovenous anastomoses (AVA) appear to play some important role in blood flow distribution. In two studies thereon using microspheres, the investigators noted that from 44 to 29% of the microspheres lodged in the lungs (Liggins et al. 1980; Zapol et al. 1979). Some of this high percentage was attributed to the aortic site of injection, which excluded flow to the brain and heart. In addition, Blix and Folkow (1983) noted that CO distribution changed in the course of a submersion of a seal. Early in the submersion up to 50% of CO went to AV shunts and 10% to the brain, but later 66% went to the brain and only 10% through AV shunts. They postulated that such shunts are necessary to maintain SV at a level at which myocardial contraction would remain effective. AV shunts in the foot web of mallards concurrent with intense general vasoconstriction during forced submersions has been observed (Djojosugito et al. 1969).

Based on the cutaneous distribution of AVA, these appear to play an important role in thermoregulation (Molyneux and Bryden 1981). In seals they are densely spread over all the skin, and are 5 to 20 times more concentrated than in sea lions. In sea lions and fur seals, the greatest concentration is in the flippers. In addition to their thermoregulatory function, they and more internal AVA could be a pathway for shunted blood flow under various circumstances including during forced submersions and during post-dive tachycardia when blood oxygenation is in progress and CO is exceeding the oxygen needs of the diver. However, it is noteworthy that there are no AVA's in the skin of the totally aquatic bottlenose dolphin, *Tursiops truncatus*, false killer whale, *Pseudorca crassidens*, or dugong, *Dugong dugon* (Molyneux and Bryden 1981).

3.2.11 Venous Circulation

During these major circulatory changes in blood distribution described above, it is probable that there is much pooling of blood within the venous system. In no group is this better studied and exemplified than in the seals. I have already referred to substantial AV shunting that occurs in which arterial blood completely bypasses organs and returns to venous reservoirs. Some detail will be given about the remarkable anatomy and function of the venous system of phocids based on the description of anatomy and circulation of the harp seal by Ronald et al. (1977).

Much of the brain drainage is through the extradural intravertebral vein (EXD) (Fig. 3.7). The internal jugular receives only a minor part of the cranial outflow. At the cranial exit the EXD is ventral to the nerve cord, but with numerous anastomoses dorsal to the nerve cord. By the time it reaches the thoracic region it has coalesced into one large vessel, which is now completely dorsal to the nerve cord, but still within the vertebral foramen, where it remains to the second caudal vertebrate (Fig. 3.7). In the course of this route to the posterior there are numerous connections to the anterior and posterior vena cava

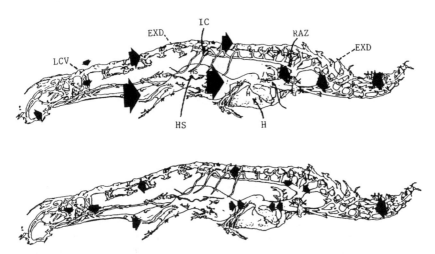

Fig. 3.7. Direction and relative velocity (*arrow size*) of venous blood flow in the harp seal while resting (*top*) and during forced submersion (*bottom*). Anterior is to the *right; AVC* anterior vena cava; *EXD* extradural intravertebral vein; *H* heart; *HS* hepatic sinus; *LCV* lumbar communicating veins; *IC* intercostal vein; *RAZ* right azygous vein. (After Ronald et al. 1977)

(PVC, AVC). The lateral intercostals link the EXD to the azygous veins. The azygous veins join shortly thereafter and the right azygous only empties into the AVC. In the lumbar-sacral region there are large, bilateral lumbar communicating veins (LCV) which join the EXD to the PVC.

At the anterior outlet of the saclike and capacious hepatic sinus (Fig. 3.7), where it passes through the diaphragm, there is a post-caval sphincter (PCS). The muscular activity of this structure is controlled by the right phrenic nerve (Harrison and Tomlinson 1963). During forced submersions the hepatic sinus expands and angiography has shown that the PCS is intermittently open and closed (Elsner et al. 1971; Ronald et al. 1977). After surfacing, flow is continuous through the sinus. Ronald et al. (1977) proposed that the two sources of blood entering the heart from the PVC and AVC have different O_2 contents and that somehow the PCS controlls the ratios (see Chap. 9, Blood Gases). For example, they noted an instant early in the dive at which there were two heart beats without blood flow from the hepatic sinus that would reduce the proportion of PVC blood entering the return flow to the heart. Presumably, later in the dive as arterial blood content falls the contribution from the PVC would be greater.

Different ratios of return of blood to the heart were observed by a method of single injection of radioisotopes as well (Murphy et al. 1980). They designated these as fast and slow circulation and estimated the fast circulation, which was presumed to be blood from the AVC, to be 10% of the total blood volume of adult Weddell seals. This injection was done 10 min after a forced submersion began, and indicates that most return flow continues to bypass the large posterior venous reservoir.

The general venous pattern in the seal shows marked alterations in rate and in the EXD, even in flow direction between rest and forced submersions. At rest, flow from the cranium through the EXD passes through the intervertebrals and intercostals to the AVC directly, or through the azygous to the AVC (Fig. 3.7). In the posterior, blood flow is anterior in the EXD up to the intercostals and then down to the azygous. The major venous return from the trunk is forward through the PVC, and flow is steady to the heart.

During forced submersion, there is no flow through the cervical intervertebrals and all flow in the EXD is posterior to the azygous, or to the intersacral veins and then forward through the PVC. The flow in the PVC is slow and pulsatile as the PCS opens and closes intermittently.

3.3 Conclusions and Summary

The discussion in the previous pages reflects some of my own research interests and experiences, but in general shows that much has been learned about circulatory anatomy and patterns in seals, followed by birds (the mallard), and reptiles, and little about sea lions and, paradoxically, almost nothing about the major and most completely adapted group for pelagic life at sea, the cetaceans. As the reader progresses through the remainder of the book the emphasis will not change. The final remarks of this chapter are based on results gleaned from experiments on seals, but with supportive evidence from the other groups, indicating that they are qualitatively correct for other divers as well. However, skepticism should be maintained while reading the above and what is to follow, because there are doubtless surprises in store concerning the mechanisms employed by some animals in diving, despite the logical ways, as we perceive them based on those of seals and ducks.

The flow patterns of forced submersion can be succinctly summarized from data obtained from microsphere studies of Weddell seals (Table 3.1) which show that in submersions of 8 to 12 min in an animal that is capable of a dive of at least 73 min (Kooyman et al. 1980) the CO drops to about 15% of pre-submersion. There is a rapid anterior circulation that incorporates about 10 to 20% of the total blood volume. Most of the known flow goes to the lung and brain. Much of the former could be due to AV shunting, which decreases as the dive progresses and brain blood flow increases. Posteriorly, or peripherally, most of the known slow circulation goes to the muscle and skin, but flow is so low that the method yields inaccurate values. The flow chart proposed accounts for only about 65 to 70% of the total during submersion, and none of the dynamics. Even in the well-studied Weddell seal there is still much uncertainty about flow patterns. How broadly interpretation about Weddell flow partitioning can be applied to other divers is uncertain, because none of the others has such a highly modified venous reservoir system. This venous system must exert a powerful influence on circulatory patterns, and blood and tissue chemistry, which suggests that other models of circulation are needed for the other groups. A further confounding

Table 3.1. Estimates of blood flow variations before and during forced submersions in Weddell seals. Pre-submersion cardiac output was $40\,l\,min^{-1}$, for which $26\,l\,min^{-1}$ were accounted. Forced submersion cardiac output was $6\,l\,min^{-1}$. Submersions were 8 to 12 min. (Zapol et al. 1979)

Cardiac output ($l\,min^{-1}$)	Pre-submersion	Submersion
MUSCLE		
Cardiac	1.0	0.14
Psoas	8.0	0.5
Intercostals	0.35	0.05
Diaphragm	0.4	0.02
VISCERAL		
Liver	0.9	0.04
Ileum	2.7	0.3
Kidney	4.0	0.36
Spleen	2.3	0.13
Pancreas	0.3	0.01
OTHER		
Brain	0.3	0.35
Bone	2.3	0.2
Fat	0	0
Skin	0.7	0.4
Bronchial	3.0	1.6

problem in whales is that the arterial circulation is highly modified due to the cervico-thoracic arterial rete system, and the great reduction of importance of the internal carotid to brain-blood supply (Galliano et al. 1966; Viamonte et al. 1968). The cetacean circulatory response to diving, which may be quite different from other vertebrates, will not be clearly understood until the mystery of the function of this retial system is solved.

In later chapters the affects of these alterations in circulation on dissolved gases, body fluids and tissues, metabolites and metabolism, during forced submersion will be discussed in comparison with changes in these variables in voluntary diving.

Chapter 4

Effect of Pressure on the Diving Animal

Several potential hazards to deep diving are a direct result of the physical gas laws of Boyle, Dalton, and Henry. Boyle's law states that at a constant temperature the gas volume varies inversely with the pressure to which the gas is subjected, $PV = C$, where P is pressure, V is volume and C is a constant. Dalton's law of partial pressure states that the pressure exerted by a mixture of gases is equal to the sum of the separate pressures that each gas would exert if it alone occupied the whole volume, thus $PV = V(P_{N_2} + P_{0_2} + P_{CO_2})$. Henry's law states that the amount of dissolved gas in a liquid at constant temperature is directly proportional to the partial pressure of the gas. Thus, $Vd = \alpha PV_L$, where Vd is volume of gas dissolved, V_L is volume of the liquid, P is gas pressure, and α is the solubility coefficient of the gas at a given temperature.

Since the rate of diffusion of a gas into the blood and tissues plays an important role in the amount of gas dissolved in a diver, it is clear that some hazards to divers using compressed air are much greater than to a breath-hold diver. In the latter case there is less total gas available to dissolve in solution, and the exposure time is usually much shorter; however, the hydrostatic pressures are often greater by an order of magnitude.

I will try to explain hazards that are of greatest risk to the natural diver by presenting a brief review of the characteristics of the various hazards found in an unnatural diver – man. Then I will discuss what is known in natural divers.

4.1 Nitrogen Narcosis

Bennett (1975) has reviewed the topic and some aspects that may be important considerations for deep-diving animals. Narcotic effects of high gas tensions were first noted in the mid-1800's, but it was about 100 years later before N_2 was shown to be the responsible agent (Behnke et al. 1935). The subject of N_2 narcosis became generally known to the layman after the invention of SCUBA, and its proliferation due to widespread use by amateur sports enthusiasts, and their interest in the underwater exploits of Jacque Cousteau. In his book "Silent World" he poetically coined the synonym for N_2 narcosis of "L' ivresse des grandes profondeurs" or "rapture of the depths" (Cousteau 1953).

The threshold for N_2 narcosis is highly variable among individuals, as well as within an individual, depending upon the level of exercise, fatigue, and tissue and blood CO_2 concentration. The first signs of narcosis usually occur at a depth of about 30 m (4 ATA, or a $Pa_{N_2} \sim 2990$ torr ~ 399 kPa), and the terminal condition of loss of consciousness occurs at about 100 m.

The causes of N_2 narcosis are complex. Bennett (1975) reviews some of the various hypotheses, commenting that the most accepted idea at the present time is that of Meyer-Overton membrane solubility, wherein there is a relationship between the affinity of gases for lipid and the narcotic effect. Some of the most potent gases are chloroform and halothane, which are highly soluble in lipids, about 1000 times more than N_2. Nitrogen falls about midway in relative lipid solubility between xenon, the most soluble ($1.7/0.067 = 2.5$ times more than N_2), and He, the least soluble ($0.067/0.015 = 4.5$ times less than N_2) of the noble gases. He does not produce narcosis at any pressure. Because of this attribute it is used for deep diving and saturation diving.

The anesthetic effect of the gases can be counterbalanced by pressure. The critical anesthetic concentration occurs when it reaches a level that causes expansion of the lipid within the membrane of neuronal cells by about 0.4%. This expansion can be offset by a pressure of about 100 ATA (Bennett 1975). A recent review of anesthetics, membrane alterations, and the effects of pressure is presented by Winter and Miller (1985). From the various observations summarized in the aforementioned reviews, it can be concluded that high pressure has an antagonistic affect on narcosis or anesthesia, which with the advent of a need for deeper and deeper dives while breathing compressed gases has become another major obstacle to deep diving; furthermore as we shall see in the discussion of high pressure nervous syndrome (HPNS), it could be a problem for deep diving natural divers.

4.2 High Pressure Nervous Syndrome

The history of this problem is brief. It was first described in the 1880's when a variety of animals under study were noted to undergo hyperactivity, spasms, and finally convulsions. Brauer (1975) reviews this history and subsequent progress. Apparently, not much more was done until the 1950's, when investigators studying the effects of deep diving noted that divers using mixed, compressed gases experienced tremors and convulsions. Some attributed the effect to the He gas used in the gas mixtures in place of N_2 (helium tremors). It has been noted also that the partial pressure of O_2 could lower the threshold for the onset of convulsions, if the O_2 concentration was $>2.5\%$ at pressures >50 ATA. This is equivalent to an alveolar O_2 tension of about 950 torr (~ 127 kPa).

Although gases of different mixtures may affect the onset threshold of HPNS, hydrostatic pressure alone is the causative agent. Brauer (1975) reviews the evidence for this in which liquid breathing mice, newts, and fish incurred HPNS in the complete absence of a gas atmosphere. Also, recall the earlier remarks on narcosis, that hydrostatic pressure antagonizes the narcotic affect of inert gases.

From several species of vertebrates that have been studied, Brauer (1975) reports values for 15 mammals, three birds and two reptiles. There is considerable variation within and among species. Mammals are the most susceptible. Man (Bennett 1975) and other primates (Brauer 1975) showed tremors at about

14 to 15 ATA and convulsions at about 50 ATA. Of the three avian species tested there were no tremors down to 40 ATA and it was 66 ATA for the two species of reptiles.

Curiously, no diving vertebrate is represented in any of these studies, or to my knowledge in later studies. Since the major cause of HPNS is hydrostatic pressure it is a matter of interest as to why those deep divers that descend to within the range of these pressure thresholds (see Table 4.1) for other vertebrates are not affected, or are they? Presumably they are not, because it is difficult to envision a shaky deep diving seal or whale at 500 to 1000 m being a very successful hunter. Furthermore, Brauer (1975) points out that compression rates of a very modest 1 m min^{-1} to a rapid 650 m min^{-1} have a negative affect on the threshold. No diving vertebrate descends at 650 m min^{-1}, but many fall in the lower range of about 100 m min^{-1} and they may do this all day or night long. It seems even more perplexing how these animals avoid the adverse nature of compression. Perhaps less intractable is an understanding of the mechanism by which they cope with the toxic influence of oxygen.

4.3 Oxygen Toxicity

The effects of breathing oxygen at high pressure were not noted until the late 1800's. Since then there have been a number of investigations determining thresholds of occurrence, symptoms, tissue damage, and the mechanisms (Wood 1975). Some of the adverse results of O_2 toxicity on tissue in the lung alone are hemorrhage, edema, and atelectasis, to mention a few. The symptoms are dypsnea, twitching muscles, and ultimately convulsions. The onset of symptoms in man occur after an exposure of about 8 h at 200 kPa, 12 h at 100 kPa, and 25 h at 80 kPa (Wood 1975). These pressures are equivalent to a $Pa_{O_2} \sim 80$ to 200 kPa. However, the onset of symptoms is highly variable among individuals and, within an individual, depend on the subject's condition and activity during exposure. In a comparative study of a diving animal, seals seemed to be more susceptible than other mammals. After about 10 min at 455 kPa of O_2, the seal experienced seizures (Kerem et al. 1972) compared to about 30 to 120 min in other mammals measured at the same pressure.

Such high oxygen tensions and long exposures as indicated above never occur in an air-breathing diving vertebrate. The dives are not long enough and they are not deep enough to raise the low O_2 concentration of the lung of a normally air-breathing animal to levels of toxicity. In measurements of a freely diving Weddell seal, the Pa_{O_2} reached only 18.7 kPa (140 torr) transiently as it descended to depth (Qvist et al. 1986). Such a low P_{O_2} occurs because oxygen is steadily consumed, reducing the already low oxygen concentration even more, and pulmonary shunts (discussed below) become functional, reducing the arterial O_2 tension. All of these conditions mean that O_2 toxicity is never a problem for a diving animal, but conversely the transiently low O_2 tensions that could result following the elevated Pa_{O_2} while the animal was at sufficient depth could be. In humans this condition is called "shallow water blackout", or syncope.

Table 4.1. Maximum depths of marine breath-hold divers. Each taxonomic group is listed according to size. H harpooned; N net, trap or cable; O direct observation; R attached recorder, T trained; S approximate sample size in which record was obtained

Species	Maximum Depth (m)	Method	S	Source
Man, *Homo sapiens*	105	T	10^2	Hong (1988)
Ridley sea turtle, *Lepidochelys olivacea*	300	O	1	Landis (1965)
Leatherback sea turtle, *Dermochelys coriacea*	475	R	11	Eckert et al. (1986)
Dermochelys coriacea	>1000	R	10^3	Eckert et al. (1989)
Common loon, *Gavia immer*	60	N	1	Schorger (1947)
Common murre, *Uria aalge*	180	N	10^4	Piatt and Nettleship (1985)
Atlantic puffin, *Fratercula arctica*	60	N	9×10^2	Piatt and Nettleship (1985)
Chinstrap penguin, *Pygoscelis antarctica*	70	R	10^3	Lishman and Croxall (1983)
Gentoo penguin, *P. papua*	100	N	1	Conroy and Twelves (1972)
King penguin, *Aptenodytes patagonicus*	240+	R	3×10^3	Kooyman et al. (1982)
Emperor penguin, *A. forsteri*	265	R	10^2	Kooyman et al. (1971 a)
Northern fur seal, *Callorhinus ursinus*	190	R	3×10^3	Kooyman et al. (1976)
South American fur seal, *Arctocephalus australis*	170	R	6×10^2	Trillmich et al. (1986)
South African fur seal, *A. pusillus*	200	R	10^2	Kooyman and Gentry (1986)
California sea lion, *Zalophus californianus*	250	T	10^2	Ridgway (1972)
Zalophus californianus	275	R	9×10^3	Feldkamp (1985)
Hooker's sea lion, *Phocarctos hookeri*	400	R	10^3	RL Gentry (pers. commun.)
Harbor seal, *Phoca vitulina*	300	R	10^3	R Delong (pers. commun.)
Phoca vitulina	600	N	1	Kolb and Norris (1982)
Weddel seal, *Leptonychotes weddellii*	600	R	10^4	Kooyman (1966)
Elephant seal, *Mirounga angustirostris*	894	R	1×10^4	LeBoeuf et al. (1988)
Common porpoise, *Delphinus delphis*	260	R	10^2	Evans (1971)
White sided dolphin, *Lagenorhynchus obliquidens*	215	T	10^2	Hall (1970)
Bottle-nose dolphin, *Tursiops truncatus*	535	T	10^2	Kanwisher and Ridgway (1983)

Table 4.1 (continued)

Species	Maximum Depth (m)	Method	S	Source
Pilot whale, *Globicephala melaena*	610	T	10^2	Bowers and Henderson (1972)
Beluga, *Delphinapterus leucas*	650	T	10^2	Ridgway et al. (1984)
Killer whale, *Orcinus orca*	260	T	10^2	Bowers and Henderson (1972)
Sperm whale, *Physeter catodon*	1140	N	10	Heezen (1957)
Fin whale, *Balaenoptera physalus*	500	H	1	Scholander (1940)

4.3.1 Diving and Extreme Hypoxia (Shallow Water Blackout)

In the early 1930's some gas exchange characteristics of the Japanese ama were reported in which end-of-dive alveolar P_{O_2}'s were sometimes as low as 3.2 kPa (24 torr) (Teruoka 1932; cited in Lanphier and Rahn 1963a). Little further experimental work was done on the subject of gas exchange in human breath-hold divers until the early 1960's. A number of reports of divers losing consciousness while breath-holding stimulated Craig (1961) to conduct a series of four experiments. The experimental protocols varied from subjects breath-holding in air at rest to hyperventilating before breath-holding and exercising. At rest, the breaking point of breath-holding (BP) on average was when the alveolar gas concentration reached a P_{CO_2} = 6.8 kPa, and the P_{O_2} = 9.7 kPa. After hyperventilation and exercise it was: P_{CO_2} = 6.5 kPa and P_{O_2} = 5.7 kPa. In a similar set of conditions, except that when the subjects were swimming the P_{O_2} decreased to as low as 4.4 kPa, the swimmer felt that he could have easily gone further. However, no further swimming distances were permitted because an alveolar P_{O_2} of 3.3 to 4.5 kPa is associated with a loss of consciousness (Craig 1961).

Soon thereafter two other reports by Lanphier and Rahn (1963 a, b) gave a detailed description of alveolar gas tensions and gas exchange during surface breath-holds and during simulated dives to depth. During surface breath-holds there is a rapid rise in alveolar P_{CO_2} to a level that late in the dive arterial and venous P_{CO_2} are equal and CO_2 exchange ceases. The early increase in lung P_{CO_2} is due to the concentrating affect of O_2 uptake, Haldane effect of the blood, and shrinking of the lungs (Lanphier and Rahn 1963 b).

When the diver is compressed in a simulated 60-s dive to a depth of 10 m, pulmonary gas tensions and exchange follow a pattern of steady decline in O_2 content of the lung and little change in CO_2 under mild work conditions (Fig. 4.1A, Fig. 4.1B). The pattern of O_2 exchange was comparable in both the surface dive and the dive to depth, but that of CO_2 exchange was not. Lung CO_2 content steadily increased during most of the surface dive, but during a dive to depth it decreased during descent and at depth. A CO_2 gradient reversal occurred and is

33

understood when alveolar P_{CO_2} is noted (Fig. 4.1B). During compression the P_{CO_2} rises rapidly to a high level from a starting $P_{CO_2} \sim 3.3$ kPa and goes to $P_{CO_2} \sim 8$ kPa. There is also a marked rise in alveolar P_{O_2} upon descent, and a later decline to about 5.3 kPa at the end of the dive.

After hyperventilation followed by compression and mild work, the alveolar P_{O_2} and P_{CO_2} patterns are similar to the shorter dives, except for the low P_{O_2} of 3.2 kPa at the end of the dive, and the decline in O_2 uptake and possible reversal of O_2 exchange during ascent. Oxygen uptake was calculated on the basis of O_2 concentrations measured, and previously determined residual volumes and vital capacity. Therefore, for the two specific dives measured a slight error in any of these values would yield the apparent reversal. Nevertheless, it is clear that O_2 uptake in either direction is at a very low level upon ascent and the lung is no longer a useful O_2 store.

Similar results were noted when divers actually descended and ascended to and from a depth of 27 m (Schaefer and Carey 1962). During the descent, CO_2

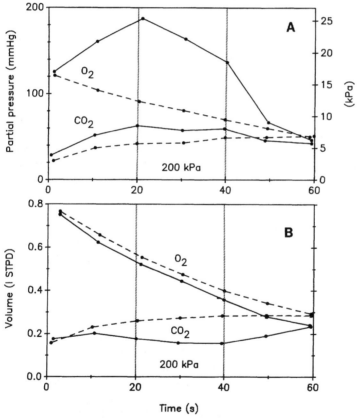

Fig. 4.1. A Alveolar P_{O_2} and P_{CO_2} in man during breath-holding and mild work at the surface (*dashed line*) and during simulated dives to 10 m (200 kPa) (*solid line*). *Hatched lines* indicate the time at depth. (After Lanphier and Rahn 1963 b). B Volumes of O_2 and CO_2 (corrected to standard conditions) in the lung of man during breath-holding at the surface and at a simulated 10-m depth. Line symbols are the same as in A. (After Lanphier and Rahn 1963 a)

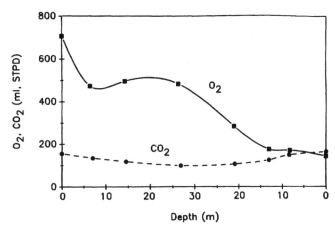

Fig. 4.2. CO_2 and O_2 volumes of the lung of man during dives to 27 m (2735 kPa). (After Schaefer and Carey 1962)

content in the lungs steadily declined (Fig. 4.2), indicating a reversed CO_2 gradient in which CO_2 was diffusing from the lung to the blood due to the hydrostatic pressure, inducing a high P_{CO_2}. The gradient reversed, or returned to a normal diffusion direction upon ascent. The authors report no change in direction of O_2 diffusion, although at times there was a very low uptake of O_2. However, the method of analysis may have lacked the sensitivity to detect any O_2 gradient reversals because according to Fig. 4.2 a reversal occurred during descent! This is discordant with the previous discussion. No comment was made about the anomaly. Towards the end of the ascent O_2 uptake was again low and alveolar P_{O_2} in one instance was about 4 kPa. There was also one instance of a diver briefly losing consciousness just as he reached the surface. The first breath restored his awareness.

The pattern of alveolar P_{CO_2} and P_{O_2} changes as a diver descends to 5 and 10 m in the experiments of Craig and Harley (1968) were similar to the previous studies. The P_{CO_2} rose steadily (5 m), or was constant (10 m) while P_{O_2} steadily declined. Upon ascent, when there was a precipitous drop in Pa_{O_2}, the subjective impressions of the divers were reported (Craig and Harley 1968). Immediately after descent the divers felt near their breakpoint (BP), but this subsided until later in the dive. During ascent there was relief from the BP, which suggested to the authors that there was not only the influence of lung volume and the level of P_{CO_2}, but also the rate of change of P_{CO_2} as well. Hence, this would help to account for the BP sensation towards the end of descent followed by relief from BP until later. Upon ascent there is relief from BP because of the expanding lung, the reduced P_{CO_2} and the rapid decline in P_{CO_2}. However, these are antagonized by the rapid decline to a low level of P_{O_2} since, as Lanpier and Rahn (1963 a) noted with their subjects, this relief from the BP upon ascent was soon replaced by loss of coordination at exceptionally low alveolar P_{O_2}.

The various components leading to shallow water blackout (SWB) for human divers are summarized in Fig. 4.3 in which alveolar gas tensions are assumed to be equal (CO_2) or close to ($O_2 > 0.7$ to 1.3 kPa) arterial tensions. A diver is

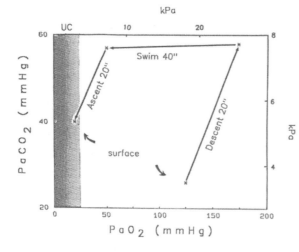

kPa

Fig. 4.3. Effect of hyperventilation on alveolar gas tensions in man during a dive to 10 m. *Shaded area* is where unconsciousness (*UC*) is expected due to low arterial O_2 tension. (After Mithoefer 1965)

predisposed to SWB by hyperventilation before the dive by reducing PA_{CO_2} to as low as 3.3 kPa (25 torr). As the diver descends, both alveolar P_{CO_2} and P_{O_2} rise, lung volume decreases, and CO_2 gradient reverses, but BP is not reached. While at depth LV decreases slowly, Pa_{O_2} declines, and PA_{CO_2} remains nearly constant. Upon ascent lung volume increases, there is a gradient reversal for CO_2 again, and very possibly O_2 gradient reverses also as the lung expands. The last, which is especially likely if the dive is very deep and the ascent rapid, will happen as the diver ascends past the last 10 m, above which the greatest expansion of the lung occurs as it doubles its volume. The O_2 gradient reversal will greatly enhance the potential of Pa_{O_2} falling below the arterial tension at which unconsciousness ensues.

How these patterns of gas exchange may affect natural divers which dive to greater depths, and for longer times, is discussed later in this chapter, where blood gas tensions of natural divers during deep dives are reviewed. Before that and in the following paragraphs, the ultimate gas tension problem of deep diving is discussed, that of most common concern in regard to deep diving, i.e., N_2 absorption, and the avoidance of gas bubble formation during decompression.

4.4 Decompression Sickness

This hazard of deep diving, also called "bends" or "caisson disease", is generally associated with compressed air breathing. Later I will show that it may be a problem of breath-hold diving as well. For now, I review briefly the general nature of the illness, the details of which have been extensively reviewed by Elliott and Hallenbeck (1975).

Observations of bubbles forming in animals subsequently exposed to lower pressures were reported as early as 1670 (Boyle cited in Elliott and Hallenbeck 1975). Bubble formation and decompression sickness became an increasingly important issue when compressed air breathing at high ambient pressures became frequent in the 1800's, to perform various tasks.

Some of the major effects of bubble formation are the denaturation of blood proteins, cell clumping, activation of coagulating factors such as Hageman's factor, which incidentally is absent in some cetaceans (Robinson et al. 1969), and disruption of tissue and formation of lipid emboli, to mention a few (Elliott and Hallenbeck 1975).

One major cause of tissue and vascular bubble formation is the elevated N_2 tension in these tissues after decompression. For example, it has been shown in cats that when they are decompressed to 101 kPa (1ATA) after being exposed to high pressure, bubble formation may occur in blood and tissues if their tension is still about 330 kPa.

Bubbles are especially likely to form in fatty tissue, and the veins draining these areas (Elliott and Hallenbeck 1975). A clear example of venous bubble formation was demonstrated by Spencer and Campbell (1968). Using Doppler and electromagnetic flowmeters, which are sensitive to the presence of circulating bubbles, they placed these devices on the inferior vena cava and the descending aorta of sheep. The sheep were first exposed to 1 h at a pressure equal to 60 m water depth, and then decompressed at a rate of 15 m min^{-1}. The initial indication of bubbles was in the inferior vena cava between 29 to 25 m. These increased from an occasional signal to a saturated output. At from 2 to 12 min after reaching the surface, arterial bubbles appeared which heralded the onset of convulsions and collapse of the animal. Possibly the lung's capacity to filter bubbles was overwhelmed and bubbles somehow passed to the arterial side of circulation. Such observations are especially important because most human work on decompression procedures is based on symptoms which no doubt do not reflect the first occurrence of bubbles. These symptom-free "silent bubbles" were earlier postulated by Behnke (1945).

The above brief reviews of decompression sickness and some other hazards of diving to depths set the stage for the following discussions on deep diving in natural divers. One hazard, oxygen toxicity, I have already dismissed as not a potential hazard to breath-hold divers based on Pa_{O_2} of diving animals (see Chap. 9). Others may seem to be especially imposing problems in light of the impressive depth capacities of some divers.

4.5 Effect of Pressure on the Diving Animal

Diving animals are exposed to a host of potentially hazardous conditions when they dive to depth. There are no experimental data to help in understanding how deep-diving animals avoid or adapt to pressure effects that, for example, are known to induce N_2 narcosis, O_2 toxicity, and high pressure nervous syndrome.

I have already dismissed O_2 toxicity because O_2 tension during a breath-hold dive can never become high enough and long enough to be a liability. When I discuss blood gas tensions it will be apparent that N_2 tensions probably do not reach levels that are narcotic in diving mammals. Blood N_2 tensions become higher in reptiles and birds, but nothing is known about the susceptibility of these vertebrate groups to narcosis. Perhaps these less cerebral animals have less of a problem with the hallucinatory effects of N_2 at high tension. As for HPNS, this is a problem that many of my colleagues have asked me about when their attention is drawn to the remarkable depth capacities of diving vertebrates, but as I commented earlier, no one has investigated the problem.

The focus of the remainder of this chapter will be on the more obvious and what I have found to be the most frequently asked questions about deep diving: First, how do the animals withstand the crushing effects of pressure when they dive to depths at which the ambient pressure may be tens or even hundreds of atmospheres? Secondly, why do they not suffer from bends? Finally, do the effects of pressure set some limits on diving behavior? The first two questions will be discussed in this chapter as well as some less considered problems of deep diving. I hope to show the reader that deep-diving animals have almost unlimited capacity to tolerate the mechanical distortions caused by compression, that "bends" are generally avoided by a great reduction in gas exchange between the lungs and blood, and/or by short-term exposure.

Perhaps the best topic to begin a discussion of "Effect of Pressure on the Diving Animal" is to summarize the maximum diving depths reported (Table 4.1) for 23 different species that range in size from the 1-kg murre to the 60-metric ton fin whale. The depth measurements have been obtained in a variety of ways. Some represent large sample sizes due to detailed behavioral studies, i.e., leatherback sea turtle, king penguin, Weddell seal, while others may be only single observations, i.e., Ridley sea turtle, harbor seal, sperm whale. As a consequence, the confidence limits of these data as estimates of operational depths of diving vary greatly.

From Table 4.1 it is clear that capabilities of different species of divers are not equal. There is a graded capability. Free-diving man looks impressive with a record of 100 m by Jacques Mayol, but this represents an almost superhuman effort. Few people, even good swimmers, can reach 10 m in free diving. Little is known about sea turtles and sea snakes but I suspect that the record of >1000 m in the leatherback sea turtle is a hint that deep diving may be common in some species. Incidentally, the 290 m depth reported by Landis (1965) for the green sea turtle is a misidentification. Prof. J. Curray showed me some of the original photographs taken of this animal, which my colleagues Scott and Karen Eckert have identified as an olive Ridley sea turtle, *Lepidochelys olivacea*.

Birds in general are relatively shallow divers, with the exception of king and emperor penguins, which are comparable to fur seals and sea lions. The depth range of fur seals appears to cluster around 200 to 300 m. However, most of these measurements were from females. There is sexual dimorphism in the group; the larger males probably dive much deeper. Seals tend to dive deeper than the fur seals and sea lions. The maximum depth records of seals range from 300 to 900 m. Once again, sexual dimorphism may make a difference. The deepest dive for a phocid of 900 m was obtained from a female northern elephant seal (Le Boeuf et al. 1988). The much larger males probably dive deeper.

There are 78 species of whales, which have the greatest range in size, shape, and behavior of all aquatic groups of diving vertebrates. There is a full range of diving depths from the shallow-diving river dolphins to the sperm whales that feed over the continental slopes at depths down to perhaps 2500 m. The depths reported in Table 4.1 highlight the need for special adaptations in the anatomy and physiology of diving animals. Boyle's law comes immediately to mind because divers have a large gas-filled space, the respiratory system.

4.5.1 Mechanical Effects

In the nondiving mammal – man – the mechanical effects of compression on the lung can be serious and depth-limiting. Craig (1968), who gives a detailed discussion of the problem, shows in a clever experiment, in which the diver precedes the dive descent by a maximum expiration, that at about 4.8 m depth 600 ml of blood have shifted into the thoracic cavity. The blood shift maintains a pressure equivalent between ambient and esophageal pressure. He then calculates from these data the actual depth a diver could reach if he dived on full inspiration (Fig. 4.4). The estimate was close to the record breath-hold at that time.

The record at that time (1968) was held by R. Croft, who had reached a depth of 73 m. It was noted that Croft had an unusual lung volume and lung capacity (Schaefer et al. 1968). Although his weight was only 79 kg, his total lung capacity (TLC) was 9.1 l and his vital capacity (VC) = 7.8 l. Thus, the calculated depth to which his lung volume after a full inspiration would equal his residual volume (RV) at depth was 60 m. Croft's unusual lung dimensions, which may have given him some of his ability to dive to great depth, are in contrast to the more normal dimensions of J. Mayol, which were TLC = 7.2 l, and VC = 5.3 l. A few years later Mayol set the present record of 101 m. In the case of both men, more than 1 l of blood would have had to shift into the thoracic cavity to maintain a balance in pressure between ambient and the thorax. Such a shift was measured in Croft when he descended to 27 to 40 m, where blood pooling in the thorax equaled 1047 and 850 ml, respectively.

The diving lung volume of sea turtles, penguins, seals and sea lions has been measured (Table 4.2). In mammals it is approximately 50% of the total lung capacity (Kooyman et al. 1972; Kooyman et al. 1973 a; Kooyman and Sinnett

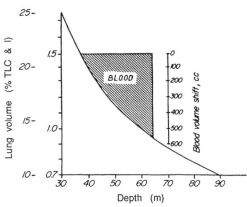

Fig. 4.4. Blood volume shift (indicated by *cross-hatching*) required for pressure equalization within the lung in a human diving to depth. This assumes maximum inspiration before the dive; gas exchange is ignored. (After Craig 1968)

1982). This reduction in volume at the beginning of the dive only slightly decreases the absolute amount gas volume at depth. For example, an animal at 100 m experiences a reduction in the initial lung volume from 5 l to 0.5 l, whereas if the initial volume were 2.5 l, it would be 0.25 l at 100 m. Either volume is beyond the limits of chest wall compliance in man, so that a large amount of intrathoracic pooling of blood is necessary to compensate for the volume change below residual volume as discussed above.

This pooling is not necessary in the seal because lung and chest compliance are unlimited (Fig. 4.5). This means that the lung and chest offer no resistance regardless of the depth to which the seal dives. The same measurements have not been made in sea lions and porpoises, so it remains conjectural that their respiratory system responds similarly. However, gas exchange in seals and sea lions during compression has been measured. The results are similar in both species, suggesting that the degree of lung and chest collapse during compression may be the same (see below).

Table 4.2. Diving gas volume relative to body weight (ml·kg^{-1}). Gas volumes are STPD

Species	Gas volume	Reference
Homo sapiens	70	[a]
Chelonia mydas	115	Berkson (1967)
Pelamis platurus	145	Kooyman (unpubl.)
Pygoscelis papua	160	Kooyman et al. (1973 b)
Mirounga angustirostris	20	Kooyman et al. (1972)
Phoca vitulina	26–32	Kooyman and Sinnett (1982)
Leptonychotes weddellii	22	Kooyman et al. (1971 b)
Zalophus califoranus	39	Kooyman and Sinnett (1982)
Enhydra lutris	330, 189[b]	Kooyman (unpubl.)

[a] Assumed a 70 kg man with a 6-l total lung capacity.
[b] Usually would dive with larger value and exhale underwater. Those that dived at lower gas volume did not exhale. Measured with similar procedures to the sea lion and harbor seal.

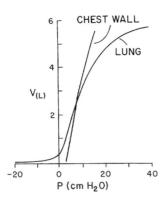

Fig. 4.5. Volume-pressure curve of the chest wall and lung of an anesthetized ribbon seal. (After Leith 1976)

Another similarity in seals and sea lions is cavernous sinuses of the middle ear (Odendahl and Poulter 1966; Graham 1967). Presumably, upon compression these sinuses engorge with blood and reduce the air space within the middle ear to an amount appropriate to maintain pressure equilibrium between the middle ear and the ambient pressure. In this way the ear can maintain its equilibrium free of the respiratory system, and the seal or sea lion need not resort to the clumsy swallowing maneuvers of humans when descending.

Mechanical effects of compression are probably minimal in diving birds because much of the gas of the pulmonary system is in air sacs outside of the thoracic cavity. The sea snakes likewise have compressible bodies so that mechanical distortions should not be a significant problem. Sea turtles, in contrast, appear to have a rigid carapace, but the plastron appears to be flexible, at least in the green and Kemp's sea turtle (*Lepidochelys kempi*). Whether it is of unlimited flexibility or sets some depth limit requires further study. Also, I noted in the cinemas of the deep-diving Ridley sea turtles, mentioned earlier that the carapace appeared to be concave on either side of the dorsal ridge.

4.5.2 Tensions of N_2 in Blood and Tissue

As an animal descends, the partial pressures of each gas will increase proportionately in accordance with Dalton's law; and I have pointed out some of the effects of partial pressure changes in the earlier discussion of shallow-water blackout. As the partial pressures increase so does the amount of each gas in solution (Henry's law). This leads us to a consideration of decompression sickness.

It might be assumed that no special adaptations are necessary for avoidance of bends in the breath-hold diver because the exposure time to high N_2 partial pressures is short and only a small amount of gas is available for absorption and the increased circulation after the dive enhances rapid, clearance of N_2 before bubble formation can occur (Irving 1935). This is perhaps true for single dives in diving birds, most species of which seldom exceed a duration of 2 min. However, calculations show that even on a single dive, a Weddell seal, a noted deep diver (Table 4.1), could absorb enough N_2 to experience high arterial and tissue N_2 tensions (Table 4.3), which Irving (1935) also conceded. I will not do further calculations on reptiles, but the case is even worse because they dive with a large gas volume and are capable of very extended dives.

N_2 tension in blood and tissue that results from a dive is dependent upon depth and duration, and distribution of the circulating blood. Blood distribution is most restricted in the forced dive in which blood flow is limited to the heart, brain and lung (Chap. 3). Under these conditions, the N_2 tension in arterial blood (Pa_{N_2}) of the elephant seal, *Mirounga angustirostris*, peaked at 300 kPa and equilibrated to 200 kPa when it was approximately the same as venous P_{N_2} (Fig. 4.6). This was true regardless of simulated depth from 37 to 138 m. All of these values are below the minimum P_{N_2} of 330 kPa found to be necessary for bubble formation in cats (Harvey et al. 1944). If blood flow distribution is more broad, such as might occur in a natural dive then the P_{N_2} blood and perfused

Table 4.3. Estimates of N_2 tension after equilibrium during dives to various minimum depths in Weddell seals. Calculations were based on the lung volume, body volume distribution, ambient pressure, gas exchange equilibrium, and dive duration sufficient to allow equilibrium among the body compartments

Distribution	Ambient pressure (kPa)	Depth (m)	N_2 Tension (kPa)
Blood	1180	>107	1060
Blood + 20% extravascular water	830	> 72	745
Total body water	415	> 31	375
Body water and fat	170	> 7	150

Assumptions: Mb = 425 kg; diving lung N_2 volume = 7.7 l (STPD); blood volume = 63 l, body fat = 128 kg, total body water = 208 l, N_2 solubility in blood = 14 ml l^{-1} 100 kPa N_2^{-1}; N_2 solubility in fat = 70 ml l^{-1} 100 kPa N_2^{-1}; alveolar N_2 fraction = 0.9. (After Kooyman 1972).

tissue would be even less than those measured during forced submersions. A few measurements of venous P_{N_2} in a 7–8-month-old Weddell seal showed values similar to elephant seals (Fig. 4.6). From such close agreement I assume that arterial P_{N_2} would also be similar.

Because of the above results it was of much interest for me to learn that during voluntary diving in sub-adult and adult Weddell seals arterial blood samples were collected in the course of dives as deep as 230 m (Falke et al. 1985). The P_{N_2} was never greater than 323 kPa (Fig. 4.7), slightly higher than our

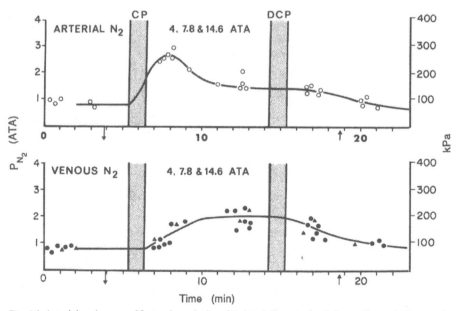

Fig. 4.6. Arterial and venous N_2 tensions during simulated dives to depth in northern elephant seals (*open and closed circles*) and a Weddell seal (*closed triangles*). Beginning and ending of dives are indicated by *arrows*. Compression and decompression periods indicated by *hatched areas*. (After Kooyman et al. 1972)

measurements in elephant seals, and this was independent of depth of dive. In fact, the deepest dive of 230 m yielded one of the lower peaks in P_{N_2}, of about 222 kPa. Incidentally, the heart rate during the 230-m dive was about equal to those found in forced dives (Zapol et al. 1979), suggesting that blood flow may have been rather restricted in distribution. It should be repeated in regard to blood distribution that forced submersions are an ideal procedure for the assessment of the hazards of inert gas exposure because the method gives the worst possible case. The very limited flow distribution will result in the highest gas tensions possible in the arterial blood, brain, and heart tissue.

Since arterial P_{N_2} no longer exceeded venous P_{N_2} at about $P_{N_2} = 152$ kPa (Kooyman et al. 1972), it might be concluded that total lung collapse, which prevented any further N_2 absorption, occurred at about 30 to 40 m. A similar value (25 to 50 m) is estimated for the free-diving Weddell seal. However, this estimate is uncertain, since venous P_{N_2} was unknown, so that it is not possible to know whether arterial and venous tensions were in equilibrium. However, for the P_{N_2} to decline while the seal is still descending (Fig. 4.8) the lung shunt must be large, and for the practical concerns of decompression sickness or nitrogen narcosis there is no longer a risk, no matter how deep the seal dives.

The importance of the diving lung volume on the maximum P_{N_2} is indicated by one elephant seal that dived with volumes equivalent to 50 ml kg^{-1} compared to the usual elephant seal value of 20 ml kg^{-1} (Table 4.2). In this animal venous P_{N_2} was as high as 525 kPa during a 270-m simulated compression dive (Fig. 4.9). Furthermore, there appear to be important species differences. Harbor seals compressed down to 200 m had arterial and venous P_{N_2} that reached as high as 500 kPa (Fig. 4.10). Upon decompression all values quickly dropped below 330 kPa so that there should be little risk of decompression sickness after a single dive. However, during the dive the P_{N_2} at times was well above the P_{N_2} of 253 kPa, which in man induces N_2 narcosis, as discussed earlier. In this matter the harbor seal remains an enigma.

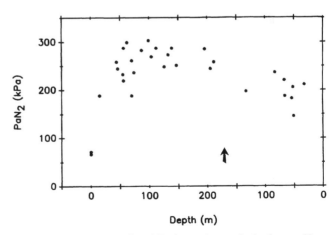

Fig. 4.7. Arterial blood N_2 tension in a free-diving Weddell seal. Each sample was obtained over a 90-s-period during a single dive. *Arrow* indicates the beginning of ascent. (After Falke et al. 1985)

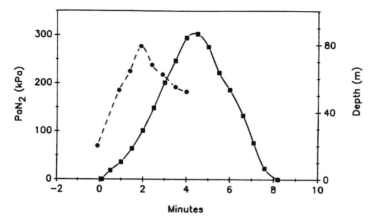

Fig. 4.8. Serial samples of arterial N_2 tension in a Weddell seal during a free dive. Each sample required 30 s for collection. (After Falke et al. 1985)

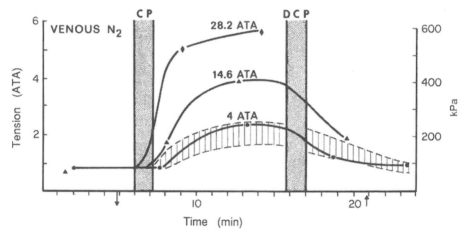

Fig. 4.9. Venous N_2 tensions of a young elephant seal during simulated dives to depth in which lung volume was unusually large. *Hatched area* is the range of venous P_{N_2} from Fig. 4.6. (After Kooyman et al. 1972)

The arterial P_{N_2} curves of harbor seals suggest that lung collapse does not occur at 30 m depth, as estimated for the elephant seal, but at a substantially greater depth even though lung volume upon diving was about the same (Table 4.2). During a compression submersion to 30 m the arterial P_{N_2} increased throughout the time at depth (Fig. 4.10). Also, compression submersions of 136 and 200 m resulted in arterial and venous N_2 tensions substantially higher than those at 30 m.

In an attempt to quantify the lung collapse relative to the degree of compression submersion, pulmonary shunts were measured in harbor seals and sea lions during simulated depths from 40 to 90 m (Kooyman and Sinnett 1982). As in previous studies of blood N_2 tensions, the results were highly variable among individual seals. At 40 to 90 m the shunt is initially up to about 50 to 70% and

44

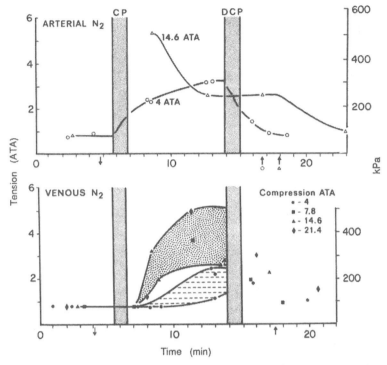

Fig. 4.10. Blood N_2 tension during simulated deep dives in harbor seals. Symbols as in Fig. 4.6. (After Kooyman et al. 1972)

steadily increases due to absorption collapse (Fig. 4.11). In both the harbor seal and sea lion there is correlation between depth and shunt up to the measured compression of 90 and 70 m (Fig. 4.12), respectively. At 70 m depth pulmonary shunt is about 50% in the sea lion. If the regression curves for the harbor seal and sea lion data are extrapolated to 100% shunt (pulmonary atelectasis) compres-

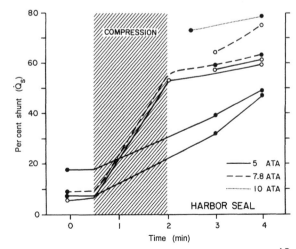

Fig. 4.11. Pulmonary shunts of harbor seals before and during simulated dives to depth. *Shaded area* is the time of compression; the period beyond is at the pressure indicated by the legend. (Kooyman and Sinnett 1982)

45

sion collapse is estimated at a depth of 170 m in the harbor seal and 160 m in the sea lion.

Estimates of total lung collapse when diving to depth were obtained in a different way for the bottlenose dolphin (Ridgway et al. 1969). In the first, and rather famous study, a dolphin was trained to dive serially to depths as great as 300 m. Mixed expired gas samples were collected at the end of the dive, and photographs of the porpoise were taken at depth.

It was noted that O_2 concentration after surface dives and dives to depth of the same duration had dissimilar O_2 concentrations. The O_2 concentration was higher in the deep dives. The authors concluded that the greater pressure should have forced more O_2 into solution. They interpreted this high O_2 concentration as the result of less consumption and isolation of O_2 from the alveoli. In addition, the unmeasured, but large amount of expirate indicated that little N_2 was absorbed. Perhaps so, but with no measured quantities of gas volumes there are other interpretations. First, the surface and deep dives are not comparable because the dolphin was aware of the depth of the target and always hyperventilated before the deep dives. It did not hyperventilate for surface dives (Ridgway pers. commun.). This difference in ventilation preparatory to the dive could account for the difference in O_2 concentration at the end of the deep dives if $\dot{V}_{O_2}$ during these dives were about the same as the surface dives. The result would be a higher lung O_2 concentration at the end of deep dives of similar duration.

Finally, the underwater photographs of the dolphin at depth show a compressed thorax in concordance with Boyle's law, but give no information about distribution of the lung gas. It could be in the alveoli or the airways. Incidentally, the conclusion that because the dolphin exhales at the end of the dive means that it dived on full inspiration is shaky. Weddell seals, which always exhale at the end of the dive, consistently dive with only 50% of TLC (Kooyman 1981). The exhalation upon surfacing seems a wise maneuver to reduce the risk of aspirating water on the first breath.

In the second study with a trained dolphin the investigators taught the animal to make a series of about 25 dives to 100 m (Ridgway and Howard 1979). Durations of dives averaged 1.5 min with surface intervals of 1 min. At the end of the series the epaxial muscle P_{N_2} was measured by means of a transcutaneous

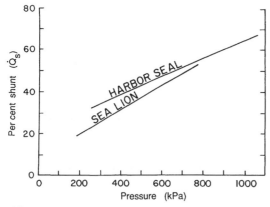

Fig. 4.12. Correlation of pulmonary physiological shunt, expressed as a percent of total pulmonary blood flow, relative to depth of dive. *Triangle* and *circle* equal the average percent shunt of the sea lion and harbor seal while resting at the surface. (After Kooyman and Sinnett, 1982)

catheter with a Teflon membrane. After about 8 min the readings were stable and P_{N_2} was about 117 kPa. This was 40 kPa above normal surface values. A curvilinear decay ratio was calculated from values obtained from 8 to 22 min after surfacing. Based upon an extrapolation of this curve to the end of the last dive, they estimated a muscle P_{N_2} of 213 kPa, making several other assumptions, such as ascent and descent rates being constant and equal, and a decay equation derived from human compressed gas breathing, they calculated a tissue half time from which P_{N_2} of the tissue should have reached before lung collapse. Their estimate was 70 m, which is necessarily a minimum depth because no account was taken of possible reduced perfusion to the muscle, nor degrees of partial lung collapse. Nevertheless, this experiment has shown for the first time that after a series of dives the P_{N_2} in the main propulsive muscle is probably well below 330 kPa, which has been shown to be a minimum tension for gas bubble formation (Harvey et al. 1944). Furthermore, for the first time analyses of serial dives was attempted.

Serial diving occurs normally when aquatic animals forage; there could be an accumulative effect of N_2 absorption. Although there is no information on natural divers, some is available for human divers. A description of the techniques used by Tuamotu pearl divers included observations on some adverse symptoms that occasionally occur (Cross 1965). Some of the symptoms were suspiciously like decompression sickness; they usually occurred after a series of dives to about 30 m. Further confirmation came from a Danish medical diving officer, who personally experienced what he diagnosed as decompression sickness after he had made a series of about 60 dives to 20 m over a 5-h period (Paulev 1965). Based upon the above, further work on serially diving marine mammals or other diving vertebrates would be of interest.

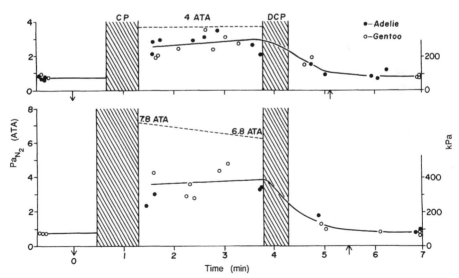

Fig. 4.13. Arterial N_2 tension changes during simulated dives to depth in Adelie and gentoo penguins. Symbols as in Fig. 4.6. *Dashed line* equals blood N_2 tension if it equilibrated with air sac N_2 pressure, assuming N_2 fraction of air sacs is 0.9. (After Kooyman et al. 1973 b)

Serial compression dives on birds or reptiles might be of even greater interest because the diving gas volumes are four to five times larger than in diving mammals (Table 4.2) (see Chap. 13). Despite such a large gas volume blood N_2 tensions in Adelie and gentoo penguins were only slightly higher than those measured in elephant seals (Kooyman et al. 1973 b). The highest values were about 455 kPa, which quickly fell to about 100 kPa upon decompression (Fig. 4.13). The N_2 tensions were only slightly higher during 68-m dives than for 30-m dives, which indicated a slowing of gas exchange at the higher pressure. It appears that in penguins, gas exchange is limiting enough at depth so that P_{N_2} does not become exceptionally high, although it borders on the level of narcosis for mammals. In contrast, high P_{N_2} may or may not occur in various reptiles; some of this is due not to compression collapse of the lung, but to a three-chambered heart. The reptilian heart has two atria and a single ventricle in which blood from the lung and systemic venous return may mix. The result when at depth would be a lower Pa_{O_2} and Pa_{N_2}, and higher Pa_{CO_2} than if the ventricle were divided.

It is a curious paradox that in aquatic reptiles, a group with less effective pulmonary gas exchange than in birds and mammals, most reports of the occurrence of decompression sickness are found. There is evidence from the vertebrae of mososaurs, extinct giant marine lizards from the Cretaceous period, that some suffered from avascular necrosis (Rothchild and Martin 1987). This disease is commonly associated with decompression exposure. In extant species some have had gas bubbles formed during single breath-hold submersion during compression experiments.

In a study of two species of sea snakes, one which is amphibious and shallow diving, *Laticauda colubrina*, and the other which is totally aquatic and probably dives to depths greater than 50 m, *Hydrophis belcheri*, it was shown that they have distinct differences in pulmonary shunt fractions and susceptibility to decompression sickness. Of total systemic flow, 28% bypasses the lung in *L. colubrina* during normal surface ventilation and 76% bypasses the lung in *H. belcheri*. With such large shunts the snakes are pre-adapted to avoid high arterial P_{N_2}, as is known to be the case in *H. belcheri*. Even at 40-m submersion depth when the shunt fraction is 83%, Pa_{O_2} was always less than 6.7 kPa and the calculated P_{N_2} would always be below the level of gas formation. Not so, however, for *L. colubrina*; despite its large pulmonary shunt at the surface it increases to only 37% at 40 m depth because the nonvascular saccular portion of the lung collapses first. As a result, P_{N_2} increases sufficiently to allow bubbles to form upon decompression, although they were not extensive enough to show adverse symptoms (Seymour 1978).

Unlike in *L. colubrina*, the entire lung and especially the anterior part is highly vascularized in the pelagic yellow-bellied sea snake, *Pelamis platurus*. When compressed, the lung gas is concentrated in the vascular section as well (Fig. 4.14), and although a large shunt of about 60% may develop (Kooyman and Sinnett unpubl.) they are additionally protected by cutaneous gas exchange. As mentioned in Chapter 2, they can remain submerged indefinitely if water P_{O_2} is maintained at 67 kPa (Kooyman and Sinnett unpubl.). The contribution of subcutaneous respiration to the total is 12% (Graham 1974). From the above it ap

Pelamis platurus

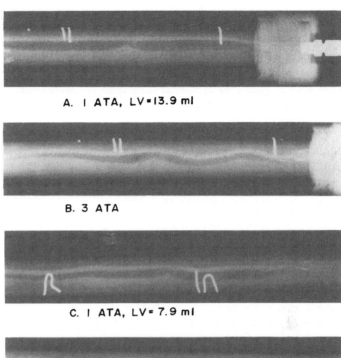

A. I ATA, LV = 13.9 ml

B. 3 ATA

C. I ATA, LV = 7.9 ml

D. 5 ATA

Fig. 4.14. Radiographs of a 43-g sea snake, *Pelamis platurus*, submerged in a compression chamber. *A* and *C* were taken when the pressure was 1 ATA, *B* and *D* after compression to 3 and 5 ATA (20 and 40 m depth). The initial lung volume in *A* was 13.9 ml (ATPS) and 7.9 ml (ATPS) in *C*. Scale can be determined from the chamber diameter, 1.9 cm. (G.L. Kooyman and E.E. Sinnett unpubl.)

pears that sea snakes are protected behaviorally, or morphologically from decompression sickness, but that under improper circumstances bubble formation can be induced after a single breath-hold dive. However, this is not the paradox I alluded to at the beginning of this discussion of reptiles.

The *only* species in which fatal decompression sickness has been induced after a single breath-hold dive, and for which blood N_2 tensions were also measured is the green sea turtle. Paradoxically, it has a three-chambered heart, so in addition to compression pulmonary shunt an intraventricular shunt could occur to offer a type of protection that birds and mammals do not enjoy. Furthermore, some turtles have a muscular pulmonary artery capable of restricting flow to the lungs (Dunlap 1955). Indeed, in the freshwater turtle, *Pseudemys scripta,* pulamonary resistance rises 150% during prolonged apnea and right to left intracardiac shunting of 60–70% may develop (Burggren 1985). This kind of anatomy would

49

seem to pre-adapt reptiles to deep diving and avoidance of "bends". However, as mentioned above, "bends" have been induced experimentally.

When a green sea turtle was submerged and compressed to 177 m the arterial P_{N_2} rose to about 800 kPa, indicating a large pulmonary shunt, but inadequate for bends protection (Fig. 4.15). At a decompression rate of 56 m min^{-1}, not an exceptionally rapid rate for diving animals, the P_{N_2} at the surface was 353 kPa. That tension is at known bubble formation levels. Several hours later the turtle died and necropsy showed numerous gas bubbles in the capillaries, cervical fascia, and heart (Berkson 1967).

A second experiment produced different results. Equilibrium between the blood and lungs was never reached, and at maximum simulated depth of 184 m there was a large differential between lung and blood P_{N_2} (Fig. 4.16). The peak P_{N_2} was 960 kPa, the highest ever recorded in a single breath-hold dive. This turtle also had bubbles form and presumably it died as well. There were two major differences between this and other experimental animals. First, blood flow must have been extremely restricted because at times heart rate was one beat every several minutes. Secondly, exposure time was much greater. The period of compression lasted up to about 70 min. This is abnormally long for a diving turtle. As described in Chapter 13, the much larger leatherback sea turtles seldom do dives for longer than 30 min, although one of the exceptionally deep dives of >1000 m lasted nearly 40 min. Finally, if and when such long compression exposures occur naturally in the diverse divers discussed above, and if gas exchange within the lung continues, both lung and arterial O_2 tensions by the end of the compression would be extremely low, and shallow water blackout (SWB) might be an important issue.

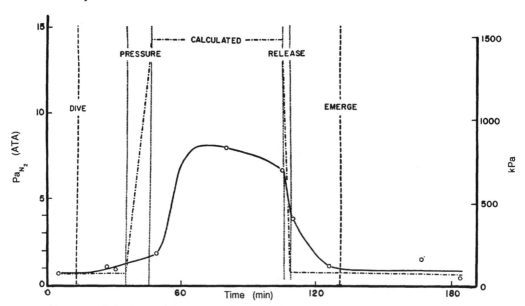

Fig. 4.15. Arterial N_2 tension in the green sea turtle during a forced submersion to a depth of 18.7 ATA (177 m). *Broken line* is the calculated equilibrium pressure of N_2 between the blood and lung. (After Berkson 1967)

50

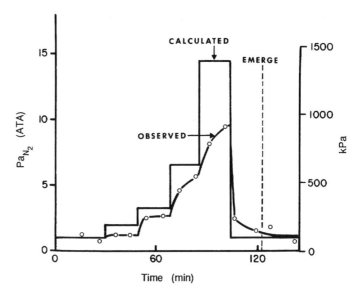

Fig. 4.16. Arterial N_2 tension during forced submersion in the green sea turtle during step increases in hydrostatic pressure up to 1964 kPa (184 m). *Calculated line* is the same as in Fig. 4.15. (After Berkson 1967)

4.5.3 Shallow Water Blackout

Perhaps the greatest risk to a deep-diving animal is not the more commonly thought about problem of decompression sickness, but brain anoxia towards the end of the dive. A substantial degree of lung collapse would be important for the avoidance of this hazard as well. It is a more acute risk. A bird or mammal, and possibly reptiles, that suffer brain anoxia for even a minute, may experience SWB. There is a risk of SWB after every dive in which a high rate of gas exchange occurs while at depth. This is due to the high partial pressure of gases that would occur in the lung. For example, at 100 m the lung partial pressure is ten times the surface P_{O_2} of 10.1 kPa. If there were no major reductions in blood and lung gas exchange due to a mechanical compression and physiological shunt within the lung, the high partial pressure of O_2 in the lung would cause rapid diffusion to the blood and subsequently raise the arterial P_{O_2}. During the time at depth, lung P_{O_2} might decrease to a low level (i.e., 5.3 kPa). As the diver ascends to the surface in 2- or 3-min expansion of the lung would result in a rapid decrease in lung P_{O_2} to perhaps 0.5 kPa ($\sim$0.5% O_2 concentration). This value would be much lower than the returning venous blood, and venous unloading of O_2 would occur. During this transient period arterial P_{O_2} might decline precipitously and become intolerably low.

For example, following a 43-min dive in a Weddell seal, the arterial P_{O_2} was still 6 kPa (45 torr) after several breaths (Kooyman et al. 1980). Such a persistent low value indicates a very low end of dive arterial and venous tension in which blood is slowly loading back to a normal surface level tension. More details on blood gas tensions are in Chapter 9.

Additional evidence from Weddell seals for low arterial blood P_{O_2} comes from the analysis of end-tidal gas tensions from the first expirate upon surfacing. The values averaged 3.6 kPa, even after short dives to depth (Wahrenbrock pers. commun.; Chap. 9). A few arterial values obtained shortly before surfacing in 10- to 30-min dives ranged from 1.3 to 4 kPa (Qvist et al. 1986). Finally, similar observations were obtained in a trained bottlenose dolphin. After a dive to 300 m of a duration of 4 min, mixed expired O_2 tension was about 2.4 kPa (Ridgway et al. 1969). Considering the size of the dolphin's vital capacity, the reduction in dead space due to the breath-hold, the movement of lung gas during compression and decompression, and the size of the sampling device, this was probably representative of end-tidal gas instead of mixed expired as the authors proposed.

It was noted in Chapter 2 that brain endpoint arterial P_{O_2} tension in seals is about 1.3 to 2 kPa. This is substantially lower than in terrestrial mammals. This may be an adaptation not so much to endure rare, prolonged dives, but rather to endure the routine, short dives to depth in which the transient high P_{O_2} during descent lowers lung P_{O_2} to such a level that during ascent Pa_{O_2} falls precipitously and the brain is exposed to short-term low P_{O_2}. More about variations in gas tensions will be discussed in Chapter 9.

4.6 Conclusions and Summary

Shallow-water blackout. This hazard seems, in my view, to be one of the most puzzling and unsolved mysteries of deep diving. The known tolerances of some divers to low arterial P_{O_2} may reflect an adaptation to this problem that deserves further study.

Decompression sickness. This condition appears not to be a problem to those mammals and sea snakes investigated, due to large compression shunts and in snakes also to intraventricular shunts. In birds, although pulmonary compression shunts occur, the major protection may be due to short exposure times and shallow depths. However, the deep and prolonged diving king and emperor penguins require investigation to understand this problem better.

Sea turtles are a mystery. They may dive to considerable depth and incur high arterial P_{N_2}, and they can develop decompression sickness on a single dive. Clearly, the problem needs to be investigated further, especially in light of the more detailed information on turtle diving behavior (see Chap. 13).

Mechanical effects of compression. Such effects apparently can be explained in seals and sea snakes, in which thoracic compliance allows adjustment to high pressure. No pertinent studies have been done on other divers; those of most pressing concern are the rigid-bodied sea turtles.

Nitrogen narcosis. This is not a likely hazard for mammalian divers because the N_2 tensions do not reach very high levels. Presumably in those groups, like turtles and penguins, in which tensions reach higher levels, there may be less susceptibility due to a higher threshold for narcosis.

Chapter 5

Oxygen Stores

Before beginning the chapters on voluntary diving the O_2 stores should be considered because they are the main factor in support of natural diving. Within this chapter I will characterize the similarities and differences between land-bound and aquatic reptiles, birds, and mammals. The O_2 stores are discussed as three compartments; the lung, blood, and muscle. This is followed by a collation into the total available body O_2 stores of five basic types of divers in which there is information on natural diving to relate to the store.

5.1 Pulmonary Storage of Oxygen

The discussion in the previous chapter showed that the lung is a better storage site for N_2 than for O_2 store, and indeed, deep divers at least should and do tend to avoid large inspirations before diving to avoid the risk of high blood N_2 tensions.

Those animals that dive with a large lung volume (birds and reptiles, Table 4.1) usually dive for short durations, or to shallow depths, but there may be exceptions. Both the king and emperor penguin, and the leatherback sea turtle consistently dive to substantial depths of 200 m or more. Do these divers have large respiratory volumes? The emperor penguin appears to dive on a large inspiration (Kooyman et al. 1971 a) so the answer would be a tentative yes until it is actually measured.

Little is known about the lung volumes of sea turtles. The relationship of body mass to lung volume in reptiles scales to 0.75 exponent (Tenney and Tenney 1970). If one estimates the intercept by solving the regression equation for it assuming a TLC of 12 l in a 100-kg green sea turtle (Tenney et al. 1974) then:

$$V = 0.379 \, Mb^{0.75}, \tag{5.1}$$

where V is TLC in l, and Mb is body mass in kg. Based on this equation the green sea turtle lung volume measured in forced submersion (Berkson 1967) was about 60% of TLC. A similar level of lung inflation and proportionally similar sized lung would give the leatherback turtle a substantial O_2 reserve of about 3.2 l, [35 × 0.6 × 0.15 (F_{O_2})]. However, recent observations by Eckert (pers. commun.) suggest that the lung is relatively much smaller in this species.

The lung volume of diving mammals has been measured in five species during forced submersions, but in only one species, the Weddell seal, during natural dives. In all five species, the harbor, elephant, and Weddell seals, and California sea otter and sea lion the lung volume is about 50 to 60% of TLC in which the TLC is based on the equation:

$$Y = 0.10 \, Mb^{0.96}, \tag{5.2}$$

where Y is volume in liters, and Mb is body mass in kg. The equation was derived originally from a collection of marine mammal lung volumes obtained in a variety of ways (Kooyman 1973). For example, some of the values are from excised lung measurements. In one series (Lenfant et al. 1970) the inflation pressure was 30% higher than from another (Tenney and Remmers 1963). Of the four species, in the latter report two are from the Sirenidae, which are a peculiar group of mammals that tend not to follow consistent characteristics of other diving mammals. The sea otter, in which lung volume is exceptionally large compared to all other mammals, was excluded because it is at the low end of the size range and exerts a powerful influence on the curve. The exceptional lung volume of the otter may be due to its probable function as a means of increasing buoyancy in an aquatic mammal that has little body fat. The extra volume would give added buoyancy to make possible the habit of floating on its back while supporting a rock, which is used for cracking shells on the chest (Kooyman 1973).

In general, it may be that none of the divers is taking full advantage of the lung as an O_2 store. However, this awaits measurements, particularly in sea turtles, birds, and dolphins, which are usually assumed to dive with lungs filled to capacity.

5.2 Oxygen Storage in Blood

Various levels of oxygen stored in blood in relation to total O_2 stores are summarized in Table 5.1. The table is not a complete survey of all published information. I have tried to obtain information on all groups of divers and particular species that are discussed in Chapters 12 and 13 on aerobic dive limits and diving behavior. There are a number of gaps to which the table draws attention. For example, there seems to be little information on diving reptiles, especially sea turtles.

5.2.1 O_2 Capacity

According to Seymour (1982), there is no statistical difference in blood O_2 capacity between reptilian divers and nondivers; in sea snakes the advantages of a high hemoglobin concentration and blood volume would not be realized in any case because the arterial blood is not more than 80% saturated. Nevertheless, more information, especially on sea turtles, would be interesting because in one sample from the leatherback sea turtle, a very active diver, the hemoglobin concentration of 12.5 g 100 ml^{-1} (P.J. Ponganis, pers. commun.) is close to that of

Table 5.1. Blood and muscle variables related to O_2 capacity

Species	Hematocrit (%)	Hemoglobin (g 100 ml⁻¹)	Oxygen capacity (ml 100 g⁻¹)	Blood volume (% body mass)	Myoglobin (g 100 g⁻¹)	Reference
Iguana iguana	32.6	8.4				Wood and Johansen (1974)
Varanus gouldi					0.33	Bennett (1973)
V. niloticus	24	7.1	9.3			Wood and Johansen (1974)
Coluber constrictor					0.18	Ruben (1976)
Alligator mississippiensis	30.0	8.2	10			Wood and Johansen (1974)
Amblyrhynchus cristata	31.5					Dawson et al. (1977)
Achrochordus javanicus	21	6.0	9.3			Johansen and Lenfant (1972)
Chrysemys scripta	25	9				Pough (1980)
Caretta caretta	29	9.8		6.6	0.29	Lutz and Bentley (1985)
Dermochelys coriacea	42.3	12.5	16.8			Frair (1977)
Turdus migratorius	44.8	17.1	18.8			Murrish (1970)
Columba livia	52.0	19.4	26.5	9.2	0.25	Lawrie (1950)
Cinclus mexicanus	45.8	17.4	19.2			Murrish (1970)
Pygoscelis adeliae	46.2	16.5	22.4	9.3	3.0	Lenfant et al. (1969 a); Weber et al. (1974); Milson et al. (1983)
P. antarctica	52.8	19.6	26.3			Milson et al. (1973)
P. papua	43.4	16.4			4.4	Weber et al. (1974); Milson et al. (1973)
Eupdyptula minor	40	18.0	24.6		2.8	Mill and Baldwin (1983)
Aptenodytes forsteri	47.0	16.9	23.1		4.3	Lenfant et al. (1969 a); PJ Ponganis (unpubl.)
A. patagonicus	53	19.8	26.2		4.3	GL Kooyman and MA Castellini (unpubl.) Baldwin et al. (1984)
Pelecanoides georgicus	62	19.7	26.5			GL Kooyman and MA Castellini (unpubl.)
Gavia stellata	54.0	20.7	27.7	13.2		Bond and Gilbert (1958)
Homo sapiens	45	14.5	20	7		
Trichechus manatus	46.6	14.8			1.5	White et al. (1976); Blessing (1972)
Enhydra lutris	48	17.1	21.5	9.1	3.1	Lenfant et al. (1970); Castellini and Somero (1981)
Odobenus rosmarus	41.9	16.2	23.4	10.6	3.0	Lenfant et al. (1970)
Zalophus californianus	49.6		23.0		3.2	Lenfant et al. (1970)
Callorhinus ursinus	48.7	17.0	19.6	10.9	3.5	Lenfant et al. (1970); Castellini and Somero (1981)
Phoca vitulina	58.2	21.1	29.3	13.2	5.5	Lenfant et al. (1970)
					4.4	Castellini and Somero (1981)
P. fasciata	66.6	24.5	34.2	13.2	8.1	Lenfant et al. (1970)
Leptonychotes weddellii	58	23.7	32.5	14.8	4.5	Kooyman (1968); Lenfant et al. (1969 b)

Table 5.1 (continued)

Species	Hematocrit (%)	Hemoglobin (g 100 ml^{-1})	Oxygen capacity (ml 100 g^{-1})	Blood volume (% body mass)	Myoglobin (g 100 g^{-1})	Reference
Mirounga angustirostris				21.7	5.1	Simpson et al. (1970); Ridgway and Johnston (1966)
M. leonina	59.3	23.3	31.2	16		Lane et al. (1972); Bryden and Lim (1969)
Delphinapterus leucas	52.4	20.6		12.7		Ridgway et al. (1984)
Tursiops truncatus	44.1	14.4	19.6	7.4	3.3	Horvath et al. (1968); Blessing and Hartschen-Niemeyer (1969); Ridway and Johnston (1966)
Phocoenoides dalli	57	20.3	26.5	14.3		Horvath et al. (1968); Ridgway and Johnston (1966)
Phocoena phocoena			24.0		3.0	Scholander (1940); Blessing (1972)
Kogia breviceps			32.4			Lenfant (1969)
Physeter catodon	52			20	5.7	Lenfant (1969); Sleet et al. (1981)

mammals. It is 50% higher than the consistent average concentration of 8.7 g 100 ml^{-1} in freshwater turtles, crocodiles, snakes, and lizards (Pough 1980).

The few examples of diving birds include those with probably the greatest breath-hold and depth capacities (Table 5.1). In none of these is there a difference between divers and the very aerobic pigeon. Furthermore, the values fall within the normal range of terrestrial mammals so that it does not appear that diving birds have an unusual blood oxygen store. Unfortunately, there are few data on blood volume. Values for the pigeon and Adelie penguin are the same.

Similarities in blood oxygen levels between terrestrial and aquatic forms persist in mammals as well. Using man as the typical terrestrial mammal, it appears that few marine mammals differ markedly in hemoglobin concentration or hematocrit (Table 5.1). Only in the Phocidae is there a significant statistical difference from terrestrial mammals (Snyder 1983). However, in addition to the phocids there are differences in some species of whales. Both the Dall's porpoise and beluga have high blood O_2-carrying capaciy, which is directly related to hemoglobin concentration and large blood volumes.

5.2.2 Blood Volume

The blood volumes of many of the mammals may be even functionally larger than is apparent if considered on a lean body weight basis rather than on total body weight which, in most diving mammals, includes an additional fat deposit of 20 to 30% of total weight. Other aspects of blood volume are pointed out in the review of "Respiratory Adaptations In Diving Mammals" (Snyder 1983). Snyder notes an interesting, but unexplained positive correlation between hemoglobin concentration and blood volume. He suggests that the dependent variable is blood volume, since plasma volume remains constant in terrestrial and aquatic mammals so that differences are due to the increased hemoglobin concentration.

5.2.3 Hemoglobin Concentration

Further, increasing the hemoglobin concentration has its liabilities. Since the mean corpuscular hemoglobin concentration in mammals is constant at about 35 to 40% (Lenfant 1969), which suggests the maximum concentration possible, an increase in blood hemoglobin concentration must involve an increase in cell volume. This incurs an increased viscosity, which may in turn reduce efficacy of O_2 transport in the cardiovascular system. Aquatic mammals may avoid some of this problem by an increase in cell volume and consequent reduction in cell numbers, thus also total cell surface area and resistance to flow. Most aquatic mammals have RBC counts of 3 to 4 $\times$ 10^6 mm^{-3} (Lenfant 1969), about half that of humans.

In addition to the correlation between hemoglobin concentration and blood volume there may be some correlation between blood O_2 store and need. Southern elephant seal pups have a hemoglobin concentration that is no different from terrestrial mammals, but before they begin their life at sea it develops into one of the highest of all mammals (Bryden and Lim 1969). Also, muskrats have a seasonal fluctuation in hemoglobin concentration that seems to be related to the dive duration. In summer, when most swimming is for surface foraging, the hemoglobin concentration is lower than in the winter, when much of the diving is under ice (Aleksuik and Frohlinger 1971). Since much of the O_2 store may be in the muscle it would have been interesting to know if Mb concentration varies with the season as well.

Short-term variation in blood hemoglobin concentration has been observed also. Some years ago blood samples collected at three different sites from Weddell seals killed for dog food showed that the Hct was in close agreement between the central arterial sample and that of the external jugular vein. The sample from the hepatic sinus, a very capacious reservoir in this species, had a much higher Hct. It was proposed that the high Hct may have been due to plasma skimming (Kooyman 1968).

Later it was noted that blood hemoglobin values from samples collected soon after dives were higher than those in the resting animal, and that the highest Hb, or Hct, were after the longest dives (Kooyman et al. 1980). Furthermore, the Hb steadily declined in the course of recovery. We reasoned that the Hb increase must have peaked sometime near the end of the dive when heart rate was elevated and circulatory dynamics promoted mixing of the venous reservoirs. More details of this pattern were obtained later when blood samples were obtained throughout the dive. It was found that this increase occurs within the first 10 to

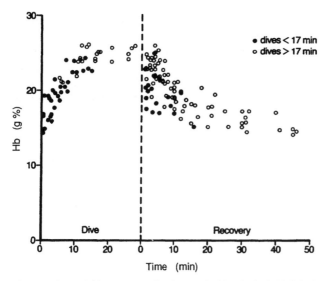

Fig. 5.1. Hemoglobin concentration in arterial blood of Weddell Seals during short dives of <17 min and extended dives >17 min and their recovery. (Qvist et al. 1986)

15 min of the dive (Fig. 5.1), and is a true increase in cell volume because plasma volume remains constant (Qvist et al. 1986).

Over a 10- to 15-min surface period the PCV returns to pre-dive levels (Kooyman et al. 1980; Qvist et al. 1986). Compared to the horse, this is a relatively rapid rate of recovery which may be due at least in part to the level of the previous exercise. After 30 min of severe exercise, the 50% increase in PCV of the thoroughbred was still as high as during the bout of exercise (Snow and Harris 1985), whereas the exercise level in the diving seal is low.

The source of these extra blood cells in the seal is unknown, but Qvist et al. (1986) have proposed that the spleen contracts and injects the cells into the circulation at the onset of the dive. This agrees with the conventional responses of some terrestrial mammals in which the PCV may increase by as much as 50% in the horse (Thomas and Fregin 1981) and 22% in the dog (Vatner et al. 1974) during vigorous exercise, and it explains the high PCV observed shortly after seals were shot. In this latter condition the stressed cardiovascular system due to hemorrhagic shock causes the spleen to contract and eject a large volume of blood cells almost directly into the hepatic sinus. These observations are all in accord with some of the earliest studies of spleen function, in which it was shown by extracorporeal procedures in the dog that the spleen contracts during hemorrhage or exercise (Barcroft and Stephens 1927), and also, by O_2 content and hemoglobin analysis that it does likewise in hypoxia (Kramer and Luft 1951). In the latter study they showed indirectly that most of the spleen's contribution must be cells alone. This is an observation that had been noted earlier by direct measurements (Barcroft and Poole 1927). Furthermore, in the horse the level of PCV increase is graded according to the level of exercise (Thomas and Fregin 1981).

58

In quest of the relative importance of the seal spleen, Qvist et al. (1986) summarized the autopsy weights of spleens from several mammals. They found that splenic weights of Weddell and southern elephant seals are about 0.9% of body weight compared to 0.2 to 0.3% for dog, man, and sheep. In another study, Fujise et al. (1985) noted an even larger proportion of 1.3 to 1.6% in two adult Weddell seals, and 0.45 to 0.95 in three neonates. It should be noted that the early report of Barcroft and Stephens (1927) was sensitive to the fact that the spleen size post-mortem is much smaller than the spleen in a resting and healthy animal. Therefore, all the above weights are only rough estimates of the functional size of the organ, which before ejection of its store is estimated by Qvist et al. (1986) to be about 30 kg or nearly 8% of a body weight of 400 kg (about the size of a newborn Weddell seal pup).

If the dive bout is considered an exercise episode, then perhaps the observations of Castellini et al (1988) that PCV rises and remains elevated throughout this period are not surprising (Fig. 5.2). Furthermore, because surface intervals are so short (2–4 min) there is not enough time for the PCV to return to resting levels. Apparently the circulating O_2 is increased to meet the needs of the active animal, only to return to low levels when the seal rests. That some margin of increase still remains is indicated by the observation that the greatest increase is observed during extended dives. This can be explained by regarding such dives as a higher degree of dive effort, such as in a sprint, compared to the routine dives.

A curious variation of this pattern of increase and decrease in PCV has been noted in elephant seals, *Mirounga angustirostris*. The PCV increases from 56 to 63% during sleeping apneusis of 4 to 11 min while resting on land (Castellini et al. 1986)! This does not seem to fit with the above-stated exercise needs.

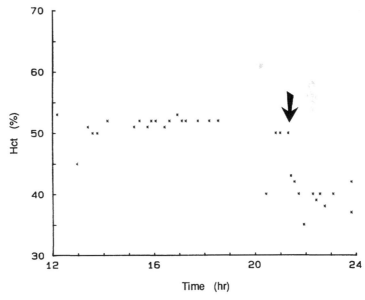

Fig. 5.2. Hematocrits of an immature Weddell seal during rest (*arrow* to h 24) and in a dive bout. Dive durations are about 5 to 10 min. (After Castellini et al. 1988)

However, during their long, pelagic sea phase elephant seals continuously spend about 90% of their time under water with surface intervals rarely exceeding 3 or 4 min (see Chap. 13). It has been suggested that even during sleep they remain submerged (Le Boeuf et al. 1985); also it was recently found that for several minutes during some dives they drift while at depths of at least 400 m (P.J. Ponganis and G.L. Kooyman, unpubl.). Are they sleeping at these great depths? If so, then it may be advantageous to extend the dive as long as possible in this condition also, hence the mobilization and increase in RBC's even while at rest under water. It will be interesting to note if other phocids that commonly sleep under water undergo a similar rise and fall in PCV. In short, at least two species of divers, both phocids, appear to sequester red blood cells when not needed, and mobilize them when diving, and the additional O_2 store is advantageous.

5.2.4 Affinity

The way blood of divers loads and unloads O_2 has been of much interest as well, and it has resulted in numerous reports on the subject. Seymour (1982), in a review of this subject in regard to reptiles, found no clear relationship to divers (I presume he means a relation to known breath-hold capacity). Instead, it may be more related to reptiles' aerobic scope, thus lower affinity in those able to attain high metabolic rates. It has been proposed also that the low affinity of reptile blood P_{50} of about 8 kPa (60 mmHg) helps compensate for the 2.5 times greater diffusion distance in reptiles compared to mammalian tissue (Pough 1980).

The role of blood O_2 affinity in mammals is no less speculative, and Snyder (1983) reviews the various proposals, which often contradict each other on the role of O_2 affinity and the Bohr effect. Snyder offers his own hypothesis that the O_2 affinity is related to the use of the lung as an O_2 store. A high affinity is most effective in those divers that dive with a large lung volume and use the O_2 store. If the affinity were low the blood would give up O_2 to the lung in respiratory and metabolic acidosis, and during ascent when the lung O_2 partial pressure is reduced. The proposal is interesting, although there are few supportive data. Some support is found in Lenfant's review (1969) of blood properties of diving mammals. He shows that in small cetaceans, which are presumed to dive on a large lung volume, the P_{50} is about 3.5 kPa (26 mmHg), and in the Pinnipedia, in which lung volumes were obtained during diving, the affinity is about 4.1 kPa (31 mmHg) (see Fig. 9.1 and Chap. 9 for further discussion).

5.3 Muscle Oxygen Store

As vague as the various properties of blood appear in regard to diving ability in aquatic animals, the muscle O_2 store relationship seems clearcut, at least in endotherms. Unfortunately, there are few myoglobin measurements for aquatic reptiles to add further comparisons. In birds and mammals the myoglobin concentration is 10 to 30 times greater in aquatic forms than in terrestrial species (Table 5.1). This is true even for the slothlike manatee.

The significance of the high muscle O_2 store to the diver, which is complex, is discussed in detail in Chapter 8 in regard to the management of O_2 stores. For now my purpose is to point out this striking difference from terrestrial forms, and to note how this enhances the total body O_2 store, and in the final analysis it may, of all the body O_2 compartments, correlate most closely with the aerobic dive durations of aquatic animals.

5.4 Total Body Oxygen Stores

Most body O_2 stores have been calculated on the basis of several assumptions. Depending on these assumptions, the O_2 store calculated may be the absolute total or the "available" store (Table 5.2).

Table 5.2. Oxygen stores of selected diving vertebrates from different groups

Species	Mass	Total O_2	Available O_2	Reference
	(kg)	(ml O_2 kg^{-1})		
Caretta caretta	20		22	Lutz and Bentley (1985)
Pygoscelis papua	6		46[a]	Kooyman and Davis (1987)
Aptenodytes patagonicus	15		58[a]	This chapter
Homo sapiens	70		26	Rahn (1963)
H. sapiens	72		19	Cross et al. (1968)
Canus familiaris	26		21	Kerem and Elsner (1973)
C. familiaris	18		15	Cross et al. (1968)
Enhydra lutris	28	52[b] (78)		Lenfant et al. (1970)
Phoca vitulina	20		40	Packer et al. (1969)
P. vitulina	24	65[a]		Lenfant et al. (1970)
P. fasciata	52	79[a]		Lenfant et al. (1970)
Leptonychotes weddellii	450	62[b] (67)	59[a]	Kooyman et al. (1983)
Mirounga angustirostris	333		73	LeBoeuf (pers. commun.)
M. leonina	333	83[c]	79	This chapter
Zalophus californianus	90		38[a]	This chapter
Callorhinus ursinus	40		39[a]	Gentry et al. (1986 b)
Tursiops truncatus	200		35[a]	This chapter

[a] These calculations are based on oxygen concentration of the lung = 15%; Mb oxygen affinity = 1.34 ml O_2 g^{-1}; Blood volume is 0.33 arterial and 0.66 venous, venous content is 7 ml O_2 100 ml less than arterial blood (Lenfant 1970) or 5 ml O_2 100 ml less than aterial (Kooyman et al. 1983; Kooyman and Davis 1987). Available O_2 store means arterial store from 95% to 20% saturated. Muscle mass is 0.30 of total body weight.
[b] Based on diving lung volume of 50% TLC. Lenfant data based on TLC, which would equal number in parentheses. Penguin lung volume is after inspiration and based on value in Table 4.2.
[c] Using the blood volume and myoglobin obtained for M. angustirostris.

Several of the values in Table 5.2 are derived from calculations in which TLC was used (Lenfant et al. 1970). Most, if not all of these species dive with 50% of TLC, which makes about a 10% difference in the total body O_2 store. Furthermore, "available" O_2 store is another 5% less than the absolute total O_2 store because the calculation of the available store considers that not all of the blood O_2 can be extracted for metabolism. Another assumption is that muscle is 30% of the total body mass. This standard value may err on the conservative side since the muscle weight of some seals (southern elephant seal and crabeater seal) ranges from 28 to 47% of total body weight (Bryden 1972; Bryden and Erickson 1972). Finally, the available O_2 store measured by 18_{O_2} dilution method (Packer et al. 1969) shows that the calculated available O_2 stores are perhaps 40% too high if, in those cases of the harbor seal, man, and dog, in which both estimates have been done, they reflect a general trend (Table 5.2). In the case of the dog, the calculated available O_2 store was based on blood gas tensions at an endpoint determined from brain wave patterns (Kerem and Elsner 1973). All other estimates have been set at an arbitrary level in which venous O_2 content at the beginning of the dive is 5 vol percent less than arterial and at the end is zero.

The calculations show that bird and mammal divers broken down into major taxa (Table 5.3) all have markedly greater body O_2 stores than man or dog. However, the distribution of the stores among the various diving groups is variable. In man the greatest store is in the lung if he dives on TLC, which is usual, but if he dives on residual lung volume it falls to 27% of total body O_2 store (Fig. 5.3). In penguins the respiratory O_2 store is also the highest, despite a high Mb concentration. In contrast, the largest volume in seals is in the mobile source, the blood, which has 64% of the total O_2 store.

The mention of blood as a mobile O_2 store raises the complex question of how the body O_2 store is managed. For example, which tissues work with their own store, and which must have supplementary stores. This issue is discussed in detail in Chapter 8.

Table 5.3. Calculated available oxygen stores of the major taxa

Family	Species	Available O_2	Lungs	Blood	Muscle
		(ml O_2 kg^{-1})	(%)		
Chelonidae	(*D. coriacea*)	20	52	42	6
Spheniscidae	(*A. patagonicus*)	58	29	38	33
Homidae[a]	(*H. sapiens*)	20	24	57	15
Phocidae	(*L. weddellii*)	60	7	65	28
Otariidae	(*C. ursinus*)	40	13	54	33
Delphinidae	(*T. truncatus*)	35	22	30	48

Calculations are based on the following assumptions: Available blood O_2 store is from arterial blood 95% saturated and extraction to 20%. Venous blood is 5 vol.% < arterial and extracted to zero. Venous fraction of blood volume is 0.66. Myoglobin O_2 affinity = 1.34 ml O_2 per g of Mb. Muscle mass is 0.3 of total body mass. Diving lung volumes of marine mammals are 0.5 of equation 5.2, and 1.0 of (5.1) for the sea turtle. Of the lung volume 12.5% is extractable O_2.
[a] Values based on Rahn (1963).

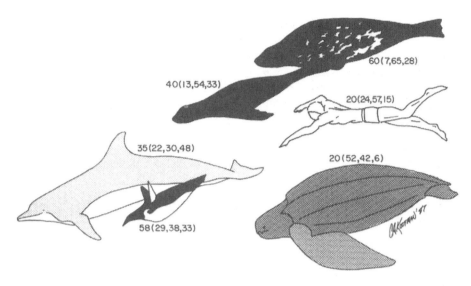

Fig. 5.3. The generalized total oxygen store of the major taxa of marine divers expressed in ml O_2 kg^{-1}. The *numbers in parentheses* are the percent of total O_2 store found in the lungs, blood and muscle, respectively. The values are derived from data in Table 5.3

5.5 Hypothesis on the Limits of Durations of Dives

I will conclude this chapter with a few comments on an unconventional proposal that the limit to dive duration is not set by depletion of O_2 stores, but rather by depletion of glucose stores. This proposal that blood glucose is the limiting factor in dive duration of the Weddell seal was proposed in a review of functions of the central organ's function during diving (Hochachka 1981). I will deal primarily with the issue of glucose versus O_2 stores being the main limit to a single, extended breath-hold dive. In particular, that "... there is extensive depletion of blood glucose during diving ..." "... blood glucose reserves are a major, if not the only source of carbon for metabolism during diving ..." (ibid p. 509) and "... skeletal muscle ... must rely fairly exclusively on anaerobic glycolysis" (ibid p. 511). How widely accepted such conclusions are now is uncertain, but they are of historical interest because they reflected accepted opinions before studies of voluntary diving were done, and a differentiation was not made between: (1) forced submersion and free dives, (2) whether serial dives or a single prolonged dive were being considered, and (3) the probable importance of fatty acids as a carbon source. These will be discussed in later chapters, and before I focus exclusively on the O_2 and glucose stores a comment on perspectives is needed.

Most of the calculations in support of the glucose hypothesis are based upon a reported 1.2-h dive by a Weddell seal (Kooyman et al. 1980). In over 15 years of study, and more than 10,000 recorded voluntary or free dives, this 73-min record is the only one that my colleagues and I have measured of such length. Indeed, there have been only about 15 to 20 dives that have exceeded 50 min, and less than 3% of all voluntary dives by Weddell seals exceed 25 min (Kooyman et al. 1980). Given these data, it seems to me that the evolutionary processes under which Weddell seals have evolved have not selected them for exceptionally prolonged dives, but rather for extending the durations of the shorter dives in which aerobic metabolism provides the energy for the dive. In this regard, the ability of the Weddell seal to make a 20–25 min aerobic dive may be exceptional; however, we lack comparative information from other marine mammals except northern elephant seals. The elephant seal has comparable or greater capacity.

I now summarize the emphasis on the importance of the O_2 store for diving. Almost every marine bird and mammal studied shows several adaptations related to increased aerobic stores. They are: (1) an extensive capacity to reduce O_2 utilization by restricting blood flow to select organs (Elsner et al. 1966; Zapol et al. 1979), some of which can return to function after an hour of no blood perfusion (Halasz et al. 1974); (2) an increase of muscle myoglobin concentration to 10 to 30 times that of terrestrial birds and mammals (Table 5.1); and in mammals (3) a tolerance to arterial oxygen tensions so low that in a human it would render them unconscious (Elsner et al. 1970 b; Kerem and Elsner 1973; Chap. 2); (4) often a large blood oxygen capacity which is a result of an unusual concentration of hemoglobin and red blood cells (Table 5.1); and (5) an exceptionally large blood volume.

If glucose were the critical metabolite for diving, then one would expect that in key organs there would be greater concentrations in diving animals compared to terrestrial mammals. Except for the large blood volume, which would mean more circulating glucose, the only enhancement of glucose reserves in divers is in the heart (Kerem et al. 1973). Blood glucose concentrations of Weddell seals are similar to other mammals. Furthermore, after dives as long as 43–61 min, blood glucose levels were not depleted. The concentrations range from 3.7 to 6.3 mM within 0.5 to 4.5 min after each dive (Chap. 8). These concentrations were little different from resting levels.

Hochachka proposed that the estimated 300 mM blood glucose store was limiting, while the blood O_2 store could last for 3 to 4 h, if it supplied only the metabolic requirements of the heart, lung, and brain. Only about 180 mM glucose are available since the store cannot be entirely used. However, other sources of glucose are present because perfusion during submersion is not completely restricted to heart, lung and brain, even during forced submersion. About 60% goes to other organs (Zapol et al. 1979). When this happens several things may occur: (1) blood O_2 is taken up, thus reducing the blood O_2 store; (2) lactic acid diffuses into the blood, which is essentially part of the glucose store, slightly degraded, (3) glucose can diffuse from peripheral stores such as the liver into the blood, thus replenishing some of the consumed blood glucose; and (4) gluconeogenesis could occur as well. Therefore, in a single, prolonged dive in which at the outset blood glucose is at normal levels, with several options to adjust for unusual glucose utilization, glucose would probably not limit the dive of a seal. If serial extended dives were made that would be another matter.

Glycogen and glucose stores are limited in seals whose diet consists mainly of fat and protein, and whose main energy store is a subcutaneous layer of blubber. It has been shown by measuring the turnover rates of labeled glucose and lactic acid before and after dives that seals use these carbon resources parsimoniously (Davis 1983). If the Weddell seals were to make serial, prolonged dives, then glucose stores might decline to such a level that they may become limiting. However, seals do not squander this resource in such a fashion.

Long recoveries and inefficient loss of diving time may explain why Weddell seals rarely make prolonged dives, although as discussed in Chapter 13, when the dives are deep the loss in bottom time due to time spent traveling to and from depth make this less clear. In all cases observed so far (see Chap. 13), the favored dive is less than 25 min, and aerobic. The selective pressures for extending the aerobic breath-hold capacities in Weddell seals and other deep divers over that of terrestrial animals and even shallow divers stem from their need for search and pursuit time at depth to obtain food. The suite of adaptations for O_2 conservation, which makes possible extended aerobic dives, carries with it pre-adaptations for a tolerance to the rare prolonged dive in which anaerobic glycolysis contributes a significant percentage of energy required for the dive.

5.6 Conclusions and Summary

This chapter has been dedicated to O_2 stores. In this, an emphasis has been placed on O_2 as the most critical resource to aerobic and prolonged natural diving. The responses discussed in the following chapters will rely greatly on this interpretation. Even in the very anoxic-tolerant turtle there is much reliance on O_2 stores and the extended anoxic episodes do not represent diving but torpor. Thus, O_2 is underwriting the diving system and most of the selection pressures have been for its enhancement and conservation, as listed in Section 5.5.

Chapter 6

Heart Rate

In this first chapter on variables observed during voluntary diving my attention
will be focused mainly on: (1) the techniques used to study heart rate, (2) the level
of change compared to resting at the surface, (3) some comparisons with forced
submersion experiments, and (4) activity. My goal is to collate, review, and bring
some order to the desultory reports on this subject. I will limit the discussion of
neurological control to a few reports necessary to help give a general concept of
the reflexive aspects of this response. Comprehensive reviews of this subject are
available (Butler 1982 a; Butler and Jones 1982; Elsner and Gooden 1983).

Although the following discussion involves a variety of reptiles, birds, and
mammals, which may have significant differences in blood volume and the fractions of cardiac output that flow to the various organs, it seems useful to mention
briefly as a starting point those values for man, a nondiver, but the most extensively studied vertebrate. About two thirds of the blood volume, at rest,
resides in veins and venous sinuses, 20% is in the arteries, 7% in the heart and
9% in the pulmonary vessels. The percent distribution of cardiac output is brain
14%, kidney 22%, liver 27%, muscle 15% and skin 6% (cool conditions)
(Guyton 1981). With this brief summary for man I will now discuss the group
most different therefrom, which also has a high degree of intragroup variability,
the reptiles.

6.1 Reptiles

In this and later chapters the reptiles discussed will include a range of types from
terrestrial lizards that take refuge in water when threatened, to the totally aquatic
sea turtles and sea snakes. Perhaps the most extensively studied reptiles are the
freshwater turtles of the genus *Pseudemys*. Appropriately, my comments begin
with a landmark paper using a member of this genus.

In 1964, when there were only five previous papers on reptilian heart rates
during forced submersions, Belkin (1964) posed the question of whether the heart
rates induced by forced submersions were different from voluntary submersions.
This was one of the first challenges to the applicability of forced submersion
results to natural diving. His subjects were 18 painted turtles, *Pseudemys concinna*, ranging in weight from 1.8 to 6.8 kg. The turtles were held at 22 °C in
25 cm of water. In 21 experiments of 8 h duration, he observed an average submersion duration of 23 min. During the 1.9% of the time spent breathing the
heart rate was 19.3 beats min^{-1} (BPM). The average rate while submerged was

8.0 BPM. Since the overall average of submersion and surface rate was 8.4 BPM, Belkin concluded that the usual rate was closest to the breath hold rate for conditions of steady state, and that the high surface rate was a ventilation tachycardia. Based upon O_2 stores, Belkin estimated that the turtle could remain submerged and aerobic for 2 to 3 h without any cardiovascular redistribution of blood flow.

Others followed with somewhat similar protocols. Four caiman, *Caiman crocodilus*, were held in a 15-cm-deep aquarium and observed by remote television (Gaunt and Gans 1969). It was noted that if someone entered the room the caiman's heart rate dropped to as low as 5 BPM, whereas, based on vaguely presented data, they concluded that resting submersions were no different from resting at the surface. As a result the authors questioned that a significant change in heart rate occurs during short voluntary dives, but speculated that there may be a significant drop at the onset of an anticipated long dive.

By using a radiotelemetry system it was possible to monitor the heart rate variation in a 45-kg *Alligator mississippiensis*, while it was free in a lake. However, the only readings reported were while the animal was close to shore and resting on the bottom at a depth of 0.5 m, and a water temperature of 27 °C (Smith et al. 1974). Submergence times were about 5 to 7 min and heart rate was between 25 to 30 BPM except when the boat approached. Then the rate fell <5 BPM. There were no comparisons between surface and submerged rates in this report.

In both of the above reports a threatening occurrence had a significant affect on heart rate. Perhaps the ultimate submergence response to threat is found in the land iguana. It is a natural trait of the land iguana, *Iguana iguana*, to take refuge in water when threatened. When it does submerge it becomes immobile and heart rate may decline to 1% of the pre-dive level, or one beat every 5 min (Belkin 1963). As an aside, it would be of interest to know what the animals' blood pressure is during this condition! Under such circumstances this lizard can tolerate submersions of up to 4.5 h (Moberly 1968). However, during unthreatened, short submersions there is no bradycardia.

If diving is considered to be an activity rather than just a resting submergence, then the above reports skirt the issue. They have noted heart rate activity during resting submersions only. In contrast to these previous reports, Heatwole et al. (1979) measured heart rate in two species of sea snakes, *Aipysurus laevis* and *Acalyptophis peronii*, after their release into a 26-m-deep coral lagoon. Heart rate was obtained with 400-m-long electrodes (the previous record for electrode length was 70 m, held by Kooyman and Campbell 1973). The heart rate during surface ventilation was 34 BPM; it was 13 BPM while resting on the bottom, and 21–42 BPM while swimming under water. While swimming, the level of heart rate appeared to be dependent upon the current and therefore swimming effort. To my knowledge this was the first observation in which underwater activity was involved.

In summary, all studies except the last mentioned have involved only resting animals in what I term resting or escape submergence to differentiate it from diving, which in my view is an active, prey-seeking process. There is more about this after the sections on mammals, but except for the sea snake study, all other reports are based upon animals that had very little behavioral variability.

Furthermore, caution should be exercised in comparisons and interpretations of active divers such as sea turtles, and wait-and-grab or stealth predators, such as alligators. Finally, in the reptiles studied only a mild decline in heart rate appears to occur between periods of ventilation. Most of these data are limited in sample size and/or vague in presentation except for the first study by Belkin (1963), in which he conducted a detailed analysis. Therefore, what the cardiovascular responses to natural diving are in reptiles, in which there is active prey-seeking, remains an open question.

6.2 Birds

There is a considerable amount of information on cardiac responses to forced submersion in birds (mainly the domestic mallard), and on the mechanisms of control. In recent years most of this work has been done by Butler and Jones, who have reviewed the subject extensively (Butler 1982 a; Butler and Jones 1982; Furilla and Jones 1986).

In a response to a broad but vague proposition that submersion bradycardias in all divers is due to stress and not representative of a dive response Blix (1985) attempted to partition the various levels of heart-rate decline to reflexogenic and psychogenic inputs. He surveyed his own investigations and those of others on forced submersions of the domestic mallard, and concluded that: (1) there is an immediate 40% reduction in heart rate, (2) after about 30–50 s a full response to 90% occurs, (3) elimination of chemoreceptor input showed that 40% of reduction is due to other inputs, (4) habituation experiments showed that 23% is emotional (others have shown about the same; 30% – Gabrielsen 1985; Gabbott and Jones 1985) and (5) that the residual 17% is due to voluntary diving of which the critical inputs are due to silencing of stretch receptors (7%), and input from the trigeminal and glossopharyngeal nerves (10%). In a somewhat comparable partitioning of responses, Jones et al. (1982) assessed only the influence of peripheral chemoreceptors after the initial and rapid decline in heart rate (Fig. 6.1). This presumably accounts for the 40% chemoreceptor input mentioned above.

The logic and applicability of the Blix report on the domestic mallard is tempered by the report of Furilla and Jones (1986), in which they compare the domestic mallard, a dabbler, with the redhead, *Aythya americana*, a diving duck. They compared the ducks while diving after hyperoxic ventilation, and after narial anesthetization. The responses of the two species were distinctly different to hyperoxia and forced submersion. No bradycardia was elicited in the dabbler, which agreed with earlier reports noted in Blix's review, but hyperoxia had no inhibition on bradycardia in the redhead. A similar result was found also in the double-crested cormorant, *Phalacrocorax auritus*, (Mangalam and Jones 1984), i.e., that bradycardia occurred despite the high arterial O_2 tension. These various results have led to the conclusion that there is little to learn from forced submersion studies about voluntary diving, as well as about divers from studying dab-

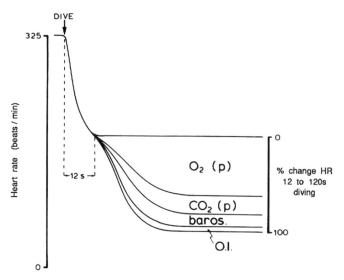

Fig. 6.1. Assessment of the contribution of peripheral chemoreceptors to bradycardia in the domestic mallard during forced submersion. The chemoreceptors were exposed to hypoxic (O_2) or hypercapnic (CO_2) blood. *Baros* stands for barostatic reflex. Unknown imputs are represented by *0.1.* (Jones et al. 1982)

bling ducks (Furilla and Jones 1986). Although this contention may not be completely warranted, it emphasizes the caution that should be exercised when trying to generalize from any forced submersion study, as well as to keep in mind the species variations that exist in the diverse divers adapted to different modes of diving.

There are relatively few reports on voluntary diving in birds; lately most of these have come from Butler's laboratory. However, one of the first studies was on a terrestrial bird from, ironically, Montana, U.S.A. The dipper, *Cinclus mexicanus*, often feeds on insect larvae in mountain streams. During forced submersions bradycardia occurred slowly, but by 15 s it was 77% of the pre-dive rate, whereas in the first 7 s of a voluntary dive to 30 cm depth heart rate was 50% of the pre-dive rate (Murrish 1970). The minimum dive rate was 33% higher than that of forced submersion, but distinctly lower than the apparent surface rate.

In a later study of the double-crested cormorant, a difference between surface and dive heart rate was not observed. Like some of the reptile studies mentioned earlier, the concept that a bradycardia occurs during natural dives was challenged as a result (Kanwisher et al. 1981). From the shallow, short dives of these cormorants, some data were presented in which rather loose comparisons were made between fast and slow surface swimming and diving. The authors stated that the heart rate was "roughly the same" under both conditions. However, the published illustration can be interpreted in other ways, dependent upon the level of activity which was not addressed.

In a series of reports on ducks and penguins, P. J. Butler dealt with the heart-rate question more quantitatively than previous authors. All experiments were

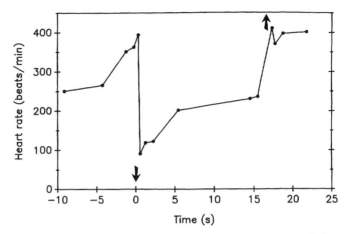

Fig. 6.2. First dives in a series of durations >12 s from tufted ducks. Sample size was 8. *Arrows* indicate start and end of dives. (After Butler and Woakes 1979)

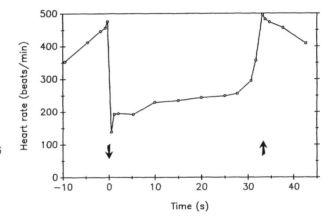

Fig. 6.3. Heart rates during 25 dives > 30 s in three tufted ducks. (After Butler 1979)

done in shallow pools ranging in depth from 0.65 to 2.7 m. Data were collected from implanted telemetric transmitters. The ducks studied were the pochard, *Aythya ferina*, and tufted duck, *A. fuligula*. Most information is from the tufted duck, in which they found that during 16-s dives heart rate dropped abruptly below a pre-dive tachycardia, and lower also than a pre-submersion heart rate (Butler and Woakes 1979). It then steadily rose, but still remained slightly below the pre-submersion heart rate (Fig. 6.2). In longer dives of 33 s duration the heart rate was stable from about 10 s until the last few seconds, when it rapidly increased (Fig. 6.3) (Butler 1979). In this study pre-submersion heart rates were much higher than the study of the same species by Butler and Woakes (1979) (cf. Figs. 6.2 and 6.3). Consequently, the bradycardia of the longer dives (33 s) appears more marked, but is actually slightly higher. If the birds were induced to dive by threat the "escape" dive heart rate fell to a slightly lower rate than during normal dives (Fig. 6.4). In addition to this psychological affect on heart rate, the influence of activity was also explored.

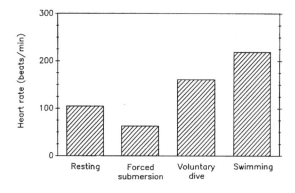

Fig. 6.4. Heart rate in tufted ducks under four conditions. Forced submersions were 415 s and on average 14.4 s during diving. Oxygen consumption for the mean dive duration and while swimming was the same and equal to 0.57 ml s^{-1}. (After Woakes and Butler 1983)

The problem was approached from an energetic aspect using special techniques to determine oxygen consumption ($\dot{M}_{O_2}$) when diving. It was shown that heart rate was consistently lower during 14-s dives, and at the same $\dot{M}_{O_2}$ as swimming at the surface (Fig. 6.4) (Woakes and Butler 1983). This suggests to me that somewhere within the body there was some degree of redistribution of and reduction in blood flow because the same $\dot{M}_{O_2}$ had a lower heart rate and, presumably, lower cardiac output. However, the authors seem not to think so from their comment that the possibility of blood flow redistribution for dives up to 25 s should be viewed with caution (Butler and Woakes 1979).

In a later paper by the same authors there was little discussion about the heart-rate variations found in three Humboldt penguins, *Spheniscus humboldti*, except that the rates were equivalent to resting on the surface of a 2.7-m-deep freshwater pool of 18 °C (Butler and Woakes 1984). The $\dot{M}_{O_2}$ of dives was 30% higher than resting $\dot{M}_{O_2}$, but the difference was not statistically significant. Again, I am impressed that similar heart rates at possibly higher metabolic rates indicate that there may have been some redistribution of blood flow during the dive.

Once again I note that, as in the reptiles, the conditions for voluntary diving in birds have been very limited. All have been in small and shallow pools with the exception of a telemetric study by Millard et al. (1973). However, the results of their study on Adelie penguins was too limited and vague in regard to heart rate to provide much insight into the problem of heart rate in natural diving. Nevertheless, an important difference from investigations on reptiles has been the added behavioral variation of exercise. As I will now discuss, this has been greatly extended in mammals in which some subjects have been trained and in others measurements during deep foraging dives have been obtained.

6.3 Mammals

All diving mammals in which heart rate has been measured during forced submersions show a profound and usually rapid onset of bradycardia, except the manatee (Fig. 3.1). A more recent study of resting submersion in the manatee also shows little decline in heart rate (Gallivan et al. 1986). With the exception of the manatee one of the only other proposals that there is no reduction in the heart rate has been based on the study of a 50-kg, restrained sea lion resting and breathing normally (Lin et al. 1972). They found the average apneusis to be 126 s, and that apnea occupied 84% of the time; consequently, they concluded that the diving bradycardia is just the elimination of the respiratory tachycardia and averages about 85 BPM. They noted that this rate was the same as that observed by Elsner et al. (1964) for a sea lion trained to dive. The rate stated by Elsner et al. (1964) in the vaguely documented report was actually 50% lower. This and most other experiments, similar to reptiles and birds, have been done in pools.

The earliest unrestrained studies were dominated by Elsner, who recognized the need for these comparisons and measured heart rates over a range from a young hippopotamus diving at will to a dolphin trained to dive. These experiments yielded some interesting and important results, but were flawed by the limited sample sizes and lack of an expression of variation in the results. In the 6-month-old hippopotamus, heart rate while diving was variable but did decrease to as low as 6% of the pre-dive value when resting. (Elsner 1966). Dives lasted about 1 min. To show the difference between forced submersions and trained animals, the results from a harbor seal trained to immerse its face on command was compared to one forced to submerge. The contrast in rates between the two conditions were illustrated in what has become a famous figure, showing that in both circumstances heart rate declined substantially, but less in the command dive (Fig. 6.5). In the young (no weight given) seal in which the trained immersion heart rate was initially 29% of a pre-dive rate of 120 to 140 BPM, it gradually fell, in the course of 7 min, to 14% of the pre-dive rate. In the forced submersion, the rate fell immediately to about 6% of pre-dive rate and remained at the level throughout the 6-min dive.

In another laboratory, training was taken to its extreme in two young sea lions (no weight given) which had implanted telemetric heart-rate monitors. One was trained to reduce heart rate, on command, to 25 BPM in the shortest time possible. Although there was a high degree of variability in their regression plot

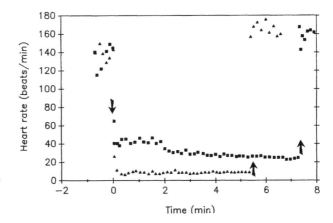

Fig. 6.5. Heart rate in the harbor seal during trained head immersion (*triangles*) and forced submersion (*squares*). (After Elsner 1965)

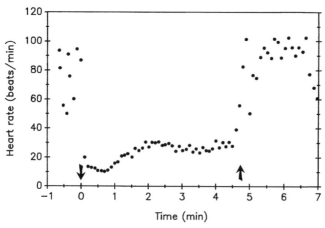

Fig. 6.6. Heart rate in a bottle-nose dolphin during a trained dive. (After Elsner et al. 1966)

(no test of significance given), the plot showed that the response time declined from about 90 s after 10 trials to 50 s after 80 trials. When one of the animals was then trained to immerse its head in a pail of water there was a rapid bradycardia, but not to as low a value as in the previous experiment. Unfortunately, little data were presented for this, or to support the comment that bradycardia in simple apnea was a much slower decline than in the trained experiments. The powerful influence of training on the sea lion and its ability to apparently "will" a change in heart rate is intriguing. Equally interesting were the results from a trained dolphin.

Dolphins are generally accepted as rather cerebral animals and perhaps as a consequence they are notoriously poor experimental subjects for laboratory submersion experiments due to a poor breath-hold response. Indeed, at times they appear to be unable to breathe after short forced submersions and although apparently conscious will asphyxiate (Scholander 1940). Yet, in a trained dolphin commanded to dive to and remain at a target 2 m below the surface for as long as 4 min 40 s, heart rate fell to an initial 17% of pre-dive rate and then steadily increased (Fig. 6.6) (Elsner 1966). It seems to me that even though the trained dives can be called psychologically forced submersions with a heart-rate response inappropriately extreme for the duration of the breath-hold, they reflect the degree of influence of higher centers, and that this response would be similar in neural control to the anticipated long or deep dive to be made by a wild animal. Therefore, the results from free animals making a variety of different dives is of special interest. So far, such studies have been done only on the Weddell seal in Antarctica.

These investigations of natural dives were done in Antarctica at McMurdo Sound, a natural marine laboratory because 2-m-thick, stable sea ice caps the Sound for an area of about 2000 km². The uniform cover, which exists from September through January, restricts the number of breathing holes and limits the distribution of the seals. This condition was used to advantage in diving studies of various kinds by cutting an ice hole in a large area of unbroken ice (Kooyman

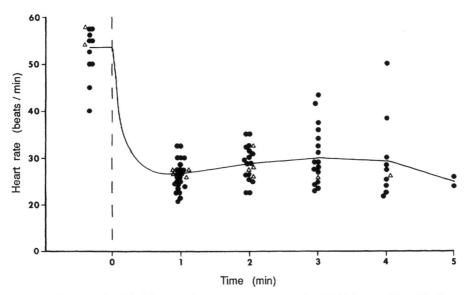

Fig. 6.7. Heart rates in adult (*closed circles*) and pup (*open triangles*) Weddell seals. The adult dives were while resting or swimming slowly under sea ice. The pups were swimming vigorously in a pool. (After Kooyman and Campbell 1973)

1965, 1968). A small, heated, laboratory was placed over the hole and seals captured elsewhere were released into the hole one at a time. Initially, the seal would make exploratory dives in seeking other holes as a means of escape from the release site. When this failed it would settle into feeding dives in which it had no competition from other seals, and in an area that had been unexploited since the winter freeze.

The first study of heart rate while diving under these conditions used 70-m-long electrode leads with breakaway connectors (Kooyman and Campbell 1973). When the seals made a hunting or exploratory dive, the first 30 s to 60 s were recorded on an ECG chart recorder. Occasionally, loafing dives near the hole were made, at which time the ECG of the entire dive was recorded (Fig. 6.7). Based on previous studies of seals in the laboratory in which the onset of bradycardia is rapid (Chap. 3), it was assumed that the heart rate at the time of lead detachment was a close approximation of the rate during most of the dive. In the report, it was proposed that there was a relationship between dive duration and heart rate in which the probability of the slope (b) not being <0 was p $=$ 0.02, using Bartlett's regression analysis. The three long dives of 28 to 50 min all had heart rates of <20, which would strongly influence the regression. If only dives of <21 min are analyzed (aerobic dives) (see Chap. 8), then the regression slope has a low level of significance, the correlation coefficient is only 0.27, and the variability is high (Fig. 6.8A). It has been corroborated recently that these short-duration recordings accurately reflect the initial heart rate of dives of short to extended durations.

Using similar capture and release procedures on Weddell seals, diving from isolated ice holes, and with a microprocessor recording unit attached, the heart rate of the entire dive was obtained (Hill et al. 1987). Heart rates initially dropped to 20 to 40 BPM, and then fluctuated between 15 and 50 BPM through the course of the dive. The average rate for the first 30 to 60 s of many dives was between 15 to 60 BPM (Fig. 6.7B)

From these investigations, it is clear that heart rates during dives of less than 21 min may be highly variable. The means ranged from about 30 to 45 BPM. This is lower than the pre- and post-dive rates of 65 to 95 BPM, which are coupled to a pre-dive hyperventilation (Kooyman et al. 1971 b), and a post-dive drive induced by altered blood chemistry (Kooyman et al. 1980). Consequently, the question arises: Is the short-duration dive heart rate a real difference from resting cardiovascular conditions and requirements?

Apparently there is not enough information for a conclusion to be drawn from the sea lion studies. Are there more definitive results from studies of seals? In past work on Weddell seals, it was shown that the resting heart rate on ice varies from 46 BPM during apneusis to 55 BPM during ventilation (Table 6.1); and when the seals are resting in water, it ranged from 35 BPM during apneusis to 64 while ventilating (Kooyman and Campbell 1973). In agreement with Belkin's (1963) premise (see above discussion), I propose that the apneustic heart rates more closely represent the steady state resting cardiac output needs of the body, and the ventilation level is a function of better ventilation/perfusion matching in the lung (discussed more later in this chapter), in which the increased cardiac rate also functions to hasten loading O_2 and unloading CO_2.

It was stated by Kooyman and Campbell (1973) that during quiet dives (Fig. 6.7) heart rate was no different from resting apneusis. However, in comparing the means of these two conditions, I found that the mean dive heart rate of 27 BPM was different from means of both the dry and wet apneustic heart rates (Student's T test $p > 0.001$ and $p < 0.001$, respectively). However, in a comparison of the quiet submersion to the active, short duration dive heart rate mean of 29, there was little difference (Student's T test 2-sided p between 0.1 to 0.2). Considering that during the dive there is active swimming, the results indicate that there was a reduced or insufficient flow to some organs. For example, recall that mammalian muscle circulation is 14% of total cardiac output. If this increases with swimming, then flow to the liver, kidney, or skin must decrease if cardiac output remains constant. This is even more apparent in the long dives (> 26 min), in which heart rates from a small sample size were consistently

Table 6.1. Heart rates of Weddell seals during various surface conditions, resting submersion, and the first 45 s of dives under sea ice. (Data from Kooyman 1968; Kooyman and Campbell 1973)

	Pre-dive	Post-dive	Sleeping in ice H_2O		Sleeping on ice	
			Eupnea	Apnea	Eupnea	Apnea
N	26	29	6	6	30	26
Mean	83	85	64	35	55	46
	Resting submersion (<5 min)		Dives (1–21 min)			
N	23		52			
Mean	27		29			

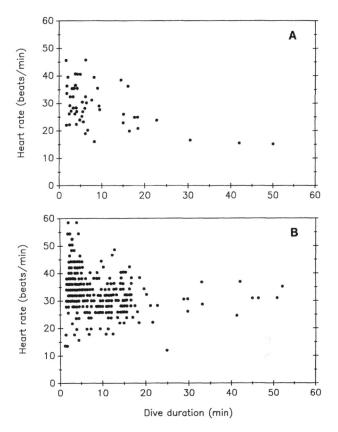

Fig. 6.8. Heart rate in relation to dive duration in adult Weddell seals. *A* Heart rate obtained in the first 45 s by 70-m break away electrodes. *B* Heart-rate averages for the first 30 to 60 s obtained by an attached microprocessor. (After Kooyman and Campbell, 1973 and Hill et al. 1987)

16 BPM (Fig. 6.8). In a much larger sample size, Hill et al. (1987) recorded average heart rates for dives >20 min of about 30 BPM. In fact, their heart rates from Weddell seals were in general lower during recovery and higher during dives.

The low heart-rate difference of long dives may also reflect a greater difference in cardiac output than is apparent from rate alone if it is recalled that stroke volume may decrease with submersion (Chap. 3). In this respect, the proposition of Jones et al. (1973), in which they proposed that if there were no reduction in heart rate an extended dive would not be possible, seems tenable. However, it seems unlikely that heart rate is the controlling factor as they state, but rather the anticipated long dive strongly influences the degree of bradycardia. This proposal agrees with the data from trained dives in which heart rate declines substantially upon command to dive (Fig. 6.5).

Furthermore, the high variability in the heart rate of shorter dives is expected. For example, low heart rates in some of the short dives (Figs. 6.7 and 6.8) may have been due to the expected possibility of a longer than usual dive, i.e., the arrival of another seal at the breathing hole and the threat that access to the surface might be hindered temporarily because Weddell seals are sometimes territorial about breathing holes (Kooyman 1968). Such an external influence is only one of several conditions that may affect cardiovascular relationships.

6.4 Neurological Control

Since no cardiovascular or cardiopulmonary reflexes have been studied during natural diving, the level of their influence is uncertain. The review of the general role of various receptors in the lung by Paintal (1973) indicates that one of the most important types in ventilatory tachycardia is the stretch receptors. Some reflexes produced by stimulation of these receptors are: (1) the Hering–Breuer reflex, in which lung inflation stimulates proportionally inhibition of inspiration, (2) increase in heart rate, (3) reduced pulmonary vascular resistance, and (4) bronchodilation.

An affect of the stretch receptors has been shown in the turtle, *Pseudemys scripta*, during forced submersions. An increase in pulmonary pressure, and presumably reduction in lung volume due to increased depth, resulted in a concomitant decrease in heart rate (Johansen et al. 1977) (Fig. 6.9). In the same report, it was shown that during submersion a stepwise reduction in lung volume caused a reduction in heart rate and pulmonary blood flow, which could be restored by injecting any of four gas mixtures: (1) air, (2) 100% N_2, (3) 100% O_2, and (4) 5% CO_2 in air.

The powerful effect of lung volume on heart rate has been assessed also in a series of well-controlled experiments on seals. When the lung of anesthetized harbor seals was inflated during forced submersions, heart rate increased, overriding enhanced bradycardia induced by stimulation of the carotid bodies (Angell-James et al. 1981). The latter was shown to play an important role in maintenance of bradycardia after the initial response due to trigeminal stimulation upon submersion (de Burgh Daly et al. 1977). The evidence for the powerful influence of the trigeminal was the difference in heart rate in anesthetized seals after apneic asphyxia, in which only the trachea was occluded, and when the face was immersed (Fig. 6.10). There was little change in heart rate during the 15-s period of occlusion. However, if occlusion were extended the heart rate eventually fell to about 30% of the pre-dive rate (Fig. 6.11; Dykes 1974 b). It fell further and more rapidly upon immersion alone, or with tracheal occlusion and immersion. It is noteworthy that heart rate of the tracheal-occluded animals, which could not exhale, was higher than in immersed animals, which did not have cannulae. So at least two, and possibly three, reflexes caused by (1) trigeminal nerve stimulation, (2) lung volume, and (3) carotid body stimulation play a role in these experiments. It was concluded that the principal sensory inputs to the submersion reflex are neural activities of the face and cessation of respiratory movements, and that the extreme regularity of the onset of immersion bradycardia is due to a reflex with little volitional control (Dykes 1974 a, b). This was summed

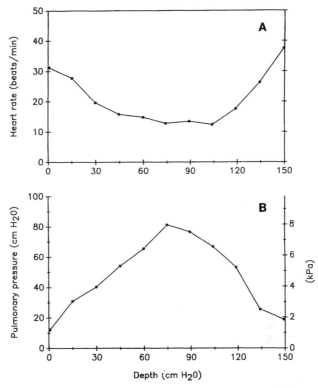

Fig. 6.9. Heart rate (*A*) and intrapulmonary pressure (*B*) induced by alterations in depth during forced submersions in *Pseudemys*. (After Johansen et al. 1977)

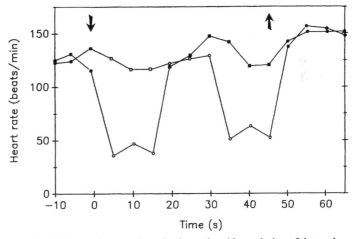

Fig. 6.10. Heart-rate response in a harbor seal to apneic asphyxia produced by occlusion of the trachea and by forced submersion. *Closed circles* air breathing; *open circles* forced submersion; *open* and *closed squares* asphyxia by tracheal occlusion. *Test 1* forced submersion followed by tracheal occlusion. *Test 2* tracheal occlusion followed by forced submersion. *Arrows* apneic period. (After de Burgh Daly et al. 1977)

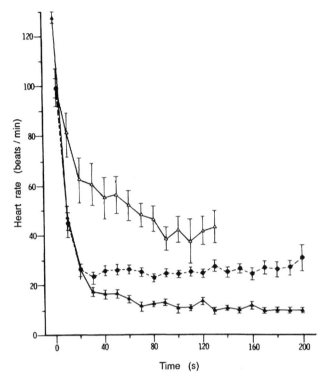

Fig. 6.11. Heart rate in harbor seals during tracheal occlusion (*top curve*), forced submersion and tracheal occlusion (*middle curve*) and forced submersion only (*bottom curve*). *Bars* are ± 1 standard error of the mean. (Dykes 1974 a)

up more specifically by de Burgh Daly et al. (1977), who stated that the initial response is due to (1) a reflex inhibition of respiration and the respiratory center due to trigeminal input that results in bradycardia because cessation of respiratory center activity releases the cardioinhibitory center, (2) excitation of the expiratory center which causes stimulation of the cardioinhibitory center; and (3) reduction in afferent vagal activity due to reduced lung volume on expiration which lessens inhibition of the cardioinhibitory center. The later response of continued low heart rate is due almost completely to carotid body influence. Evidence for this was that after a bradycardia was well established during immersion, and arterial P_{O_2} was low, the heart rate returned to the pre-dive level when the carotid body was perfused by blood with a P_{O_2} of 58 kPa (438 torr). Based upon observations of natural dives and laboratory studies of the control of heart rate, I speculate about the controlling events in a natural deep dive as a way of concluding this chapter.

6.5 Conclusions and Summary

First, that there is a considerable psychogenic influence on heart rate that power-fully influences the initial response. This sets the primary level; if it is low, very little else drives it much lower. If it is high, then as the animal submerges and descends, the wetting of the face and compression of the lung influence the rate and facilitates overriding the initial increase in Pa_{O_2} with compression (Chaps. 4, 13). With greater depth and an increasing lung shunt (Chap. 4), the Pa_{O_2} decreases rapidly and exerts an inhibitory effect on heart rate. Upon ascent, in addition to the influence of higher centers reflecting the anticipated return to the surface, the lung begins to re-expand, activating the stretch receptors and exerting its cardioaccelerator influence. Also, O_2 under pressure returning to the expanding lung from the upper respiratory tract may cause a slight rise in Pa_{O_2}, which would further contribute to an increase in heart rate. However, this latter point has not been verified by measurements of arterial O_2 tension obtained from Weddell seals (see Chap. 9).

Finally, it has no doubt been noted that nothing was said about the physical activity of the animal on heart rate. So far, a little has been published on the effects of the level of exercise in birds (Butler and Woakes 1984) and sea turtles (Butler et al. 1984). In the previously mentioned study of birds, the heart rate was correlated with $\dot{M}_{O_2}$ for swimming on the surface, but although elevated above resting on immersion, it was below that for surface swimming (Butler and Woakes 1984). Since the demand of muscle is greater, any such decrease below surface resting must reflect either insufficient flow to muscle, or increased flow to muscle and reduced flow to other organs. To my knowledge, only two studies have addressed blood flow distribution to viscera during natural dives, which will be the subject of Chapter 7.

In a study that attempted to determine the specific influence of exercise, gray seals were caused to swim in a water mill at different velocities (Fedak 1986). It was concluded that neither the surface or submersion heart rate changed with work effort, but only with duration of submersion. The consequence was a steady increase in overall heart rate as in exercising terrestrial mammals. In a similar study of harbor seals, it was found that duration of submersion decreased with workload, also. However, heart rate also increased with workload during both the surface and submersion periods (Williams et al. in prep.).

The results from all of these investigations under rather artificial conditions indicate the need for data under natural conditions, which so far are available for only one species, the Weddell seal. The emphasis has been on heart rate during submersion, but the allusions to the high pre- and post-dive heart rates have raised the question of what is the proper comparison for the dive heart rate in diving.

Earlier, I proposed it was the resting apneusis that best reflected the cardiac output needs of the body to maintain a steady-state, resting, aerobic metabolism. During these oscillations in heart rate and cardiac output, all organs except the lung fluctuate in the proportion of the cardiac output they receive. Thus, during ventilation and O_2 loading, conditions in the lung must play an important role in the stimulation of increased heart rate that results in improved ventilation/perfusion ($\dot{V}/\dot{Q}$) matching. This subject has been reviewed in detail by West (1977 a, b), where most of the data have come from human subjects. No experiments have been done on diving animals, but obviously the major function of the lung as a gas exchanger occurs during ventilation. At this time, increased blood flow, hence increased heart rate, must occur to match the ventilation. Consequently, during rhythmic apneustic periods of resting animals, the $\dot{V}/\dot{Q}$ ratio must have wide swings in its distribution from a low (<1) ratio during apneusis to good matching of about 1.0 or higher during ventilation. Thus, many of the swings in heart rate at the surface are a result of controlling reflexes regulating blood flow in lungs and other organs for better gas exchange. During some of the inter-dive ventilatory period of high output and blood O_2 loading much of the cardiac output may be flowing through AV shunts similar to that noted by Blix and Folkow (1983) during forced submersions, except that in this case it would be to insure widespread blood O_2 loading rather than to maintain cardiac pumping efficiency. The end result would be not only well-oxygenated arterial blood, but quite likely well-oxygenated venous reservoirs as well.

To re-emphasize, it is apparent from the available data that little is yet known about heart-rate variation in diving animals, particularly in reptiles and birds. The data from birds and reptiles come from limited contexts of resting submersion and the threat of danger. The submersions have been frequently referred to as "dives", but the term diving should be more restricted to an ecological perspective of an animal searching for food, socially interacting under water, or traveling from one point to another. In this light, only seals have been studied and in only two instances. However, the studies of animals conditioned to breath-holding on command are useful in obtaining a better perspective of the responses. Consequently, the conclusion, based upon existing bird and reptile data, that there is no "dive" bradycardia under natural conditions is premature and based on too limited behavioral observations (Belkin 1964; Smith et al. 1974; Kanwisher et al. 1981), or possibly even on a misinterpretation of the data (Heatwole et al. 1979). In the few mammalian studies done, there is, at times, clearly a reduction in heart rate, even when the comparisons are made with resting apneusis. Finally, because of the uncertainties about stroke volume and variations in blood flow distribution, an understanding of the meaning of heart rates is not possible without information about specific organ flow. The first and only attempts to deal with this much more difficult analysis will be discussed in Chapters 8 and 9.

Chapter 7

Splanchnic and Renal Blood Flow

To my knowledge there have been no studies on splanchnic or renal blood flow on voluntarily diving reptiles or birds. Therefore, this chapter concerns mammals in which information is the result of only two studies. Before dealing with these reports, a few comments on the basic aspects of renal and visceral blood flow follow.

In the resting mammal (human) renal and hepatic blood flow account for 50% of cardiac output. Thus, any reduction in flow to these organs in the diving

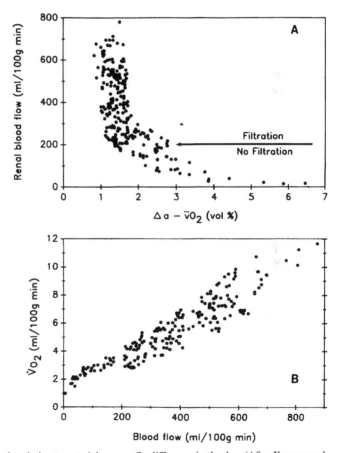

Fig. 7.1. A Renal blood flow in relation to arterial-venous O_2 difference in the dog. (After Kramer and Deetjen 1964). B Renal blood flow in relation to O_2 consumption in the dog. (After Kramer and Deetjen 1964)

animal represents a substantial savings in cardiac output and O_2 consumption. Indeed, $\dot{M}_{O_2}$ of the kidney per weight is the greatest of all organs (Folkow and Neil 1977), yet it has one of the lowest arterial/mixed venous differences of any organ (Δa-$\bar{v}$) due to the extremely high blood flow (Fig. 7.1A). Paradoxically, even though it has a very low Δa-$\bar{v}$ it behaves like a flow-limited organ because of the close relationship between $\dot{M}_{O_2}$, blood flow, Na tubular reabsorption and glomerular filtration rate (GFR) (Fig. 7.1B). If blood flow and pressure become low enough, about 20% of resting, then the kidney behaves like most other organs in being flow-independent, as $\dot{M}_{O_2}$ remains constant while flow declines. At this stage Δa-$\bar{v}$ increases and little or no filtration occurs as the $\dot{M}_{O_2}$ is sustaining basic tissue needs (Valtin 1973). It is likely that low blood flow during forced submersions serves only this basic need.

7.1 Renal Function in Diving

The reduced function of the kidney, in which the GFR nearly ceases during forced submersion of harbor seals, was described in Chapter 3. Using microsphere techniques for the determination of blood flow, it was shown, a substantiation of the older work, that renal blood flow is about 10% of resting controls during forced submersions of Weddell seals (Fig. 3.6; Zapol et al. 1979).

Consistent with the observation of a marked ability to reduce renal blood flow (RBF) is the tolerance of the kidney to ischemia and anoxia. In a series of papers it was shown that harbor seal kidney slices maintain transport function of certain organic anions and cations better than do slices from rats (Hong et al. 1982; Koschier et al. 1978). Also, excised harbor seal kidneys, after a 60-min ischemia, returned more closely to control renal blood flow and urine production levels than did similarly treated dog kidneys (Halasz et al. 1974). In this paper a bias was reflected as to the usual degree of reduction in renal blood flow by the conclusion that this ischemic resistance contributes to the ability to tolerate apneas of 25 min and normal dives of 20 min. As discussed further in Chapter 11, such long dives are probably not normal for harbor seals, and secondly, there is no evidence that RBF ceases completely for such long episodes. In fact, as mentioned earlier, RBF may remain at about 10% of resting level for tissue maintained; the results obtained by Doppler flow technique indicating no flow (Elsner et al. 1966) may mean either that the method is insensitive to such low flows, or that flow is pulsatile and easily missed. This low flow that prevails during forced submersions is probably not typical of natural dives.

The RBF was measured directly in a young sea lion trained to immerse its head on command, similar to the procedure discussed in Chapter 6. Previous to the experiment a Doppler flowmeter was placed on the renal artery. In immersions as long as 2 min, the flow velocity on average decreased by about 34% (Stone et al. 1973). However, in the illustration published the flow velocity initially fell to about that level and then rose to a rate almost equal to the value just before the pre-dive tachycardia. Somewhat similar results of small changes in RBF were shown indirectly in Weddell seals diving voluntarily.

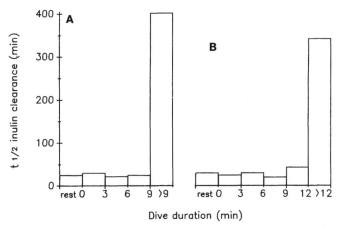

Fig. 7.2 A.,B. Relationship of half time (t 1/2) for inulin clearance at rest and during voluntary dives in seals with average body mass of 139 kg (*A*) and average body mass of 195 kg (*B*). (After Davis et al. 1983)

The conditions for the Weddell seal study were similar to those described earlier in Chapter 6. RBF was estimated in young, 130- to 205-kg diving seals by measuring the clearance rate of inulin after a single injection, and after equilibration of inulin distribution (Davis et al. 1983). Inulin, a 5200 mw polysaccharide, passes freely through the glomerular membrane, and it is not reabsorbed in the tubules. Therefore, the GFR equals the inulin clearance rate. Since RBF is closely correlated with GFR, and the filtration fraction remains relatively constant over a wide range of RBF (Bradley and Bing 1942; Schmidt-Nielsen et al. 1959), a measure of the latter gives a functional estimate of the kidney (glomerular filtration) and indirectly, RBF.

During most of the Weddell seal dives (>90%) GFR continued at resting levels, unless the duration exceeded the aerobic dive limit (ADL – Chap. 9). Then practically like an all-or-none response, GFR slowed by about 15 to 20 times (Fig. 7.2). From this study it is clear that renal function continues in naturally diving animals unless a threshold dive duration is exceeded. The threshold of the dive duration also seems to be size-dependent. More about the ADL is discussed in Chapter 9.

7.2 Splanchnic Blood Flow

Concurrent with the study of renal function in the Weddell seal was an investigation of hepatic function based on indocyanine green clearance (ICG). This dye is removed only by the liver, the flow of which in a resting human accounts for 25% of cardiac output (Guyton 1981). Blood supply to the liver has two major sources, the hepatic artery and the venous portal system. The former accounts for 25% of the flow in humans (Guyton 1981) but possibly only 2 to 3% in the resting seal (Zapol et al. 1979). The other 75% or more of hepatic blood supply comes via the spleen and pancreas (20% of portal flow), and the remainder from the gastrointestinal system. Both flow and blood volume of the liver are highly labile and blood volume may increase from a normal value of about 5 to 10% of total blood volume to three times that amount if a back pressure develops because of high resistance downstream (Guyton 1981). Such fluctuations in blood volume of the liver and hepatic sinus in seals are quite likely during some types of dives or forced submersions when the post-caval sphincter closes (see Chap. 3), causing blood pooling in the hepatic sinus and liver. The high lability of blood flow to the liver may make the liver less likely to reflect closely function or blood flow, as does inulin clearance in the kidney. However, the extraction coefficient changes only slightly from rest to exercise in man (0.77 to 0.84) (Rowell et al. 1964). If the seal is similar, then the results from the Weddell seal experiment, which show no difference in clearance from resting values, indicate little change in function when diving. ICG clearance was constant in all except one, in which clearance dropped by 25 times in a short dive of only 7 min in a 200-kg seal (Davis et al. 1983). In the only long dive (23 min) observed ICG clearance continued at the resting level. Further indirect evidence that splanchnic flow continues at some functional level was obtained rather serendipitously.

We had proposed at the conception of this study (Davis et al. 1983) that during a foraging bout of several hours digestion might be delayed until the stomach was full, thus postponing any aerobic, metabolic needs until the end of the foraging. This delay would conserve some of the body O_2 store for the neces-

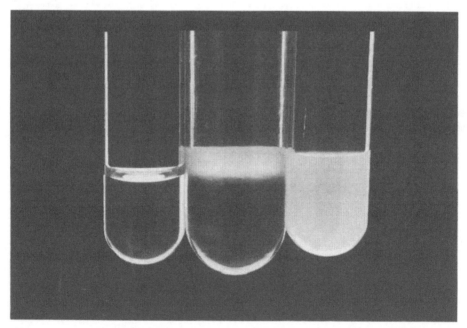

Fig. 7.3. A comparison of blood plasma from a fasting Weddell seal (*left*) and during a dive bout (*right*). Center sample is early in the bout

86

sary muscle and CNS O_2 requirements. Accordingly, we measured the stomach capacity of several seals and found that adult seals could hold about 23 l of food, enough for a substantial meal, the digestion of which could be postponed until after the foraging period. However, during the photometric analysis of the plasma for ICG, liposomes were interfering with light transmission. A comparison of pre-diving plasma clarity with plasma obtained during foraging showed a marked increase in liposomes (Fig. 7.3). Some digestion was in progress during the feeding periods and fat was being broken down and its components absorbed.

7.3 Conclusions and Summary

The analyses discussed above yield average values over the course of several dives. Superficially it seems possible that during the surface period when blood-flow and O_2 consumption are high, filtration and clearance could proceed at a rate adequate to make up for the lost time during the dive. On the occasional dive that was bracketed by samples during the pre- and post-dive period this was not indicated. It seems unlikely from the evidence presented here and later in Chapter 13 that renal and splanchnic function are on during surface ventilation and off during foraging. For example, simple arithmetic shows if the diver spends 90% of its time underwater and renal flow ceases at that time, then during the surface interval GFR must be about nine times that of resting. Since heart rates (Fig. 6.7) and cardiac output (Fig. 9.6) increase only about three to four times above resting during the surface interval, this would mean a larger part of flow would be to the kidney than seems realistic considering the need for O_2 loading of muscle and transport requirements of other organs as well.

The one measurement of specific flow in the trained sea lion (Stone et al. 1973) confirms the conclusion that the kidney is not switching off and on with diving and surfacing. Further measurements of specific flow, such as with Doppler flow meters, under natural dive conditions would resolve this uncertainty.

Chapter 8

Blood and Muscle Metabolites: Clues to Flow

In the previous chapter I discussed splanchnic blood flow, which can be determined by direct measures because the organ has discrete vascular connections, or the organ filters specific molecules such as inulin by the kidney. Furthermore, it is possible to measure turnovers of metabolites by measuring flow and concentration in the artery and vein that supply and drain the organ. Determination of flow characteristics and metabolite fluxes of skeletal muscle is a more intractable problem because blood supply is so diffuse. There has been only one direct and specific measurement of changes in muscle metabolite, i.e., decline and increase of muscle O_2 and lactate concentration [LA] in harbor seals during forced submersions (Fig. 3.5). The difficulty of the problem is reflected perhaps in the fact that the experiments, which were done over 40 years ago (Scholander et al. 1942), have not been repeated on other species as is the case of so many other experiments involving forced submersion. The most common method of analysis of muscle condition has been the indirect determination of changes in blood metabolites. As will be seen later in this chapter, that is the basis for the conclusions about blood flow and fuel consumption in natural dives.

This chapter deals with changes in blood concentrations of glucose and LA. Much of the reliance on LA concentrations is because this compound expresses the previous metabolic history of the dive, and it is most influenced by the skeletal muscles. The outpouring of LA from the muscles and other organs into the

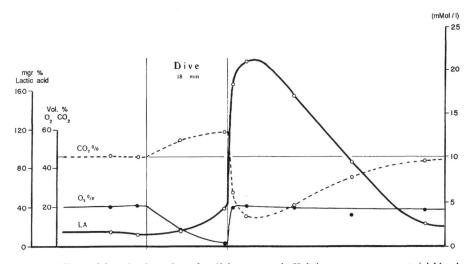

Fig. 8.1. Effects of forced submersion of a 42-kg gray seal, *Halichoerus grypus*, on arterial blood chemistry. (After Scholander 1940)

circulation after forced submersions is typified by the experiment of Scholander (1940; Fig. 8.1). The importance of those results was so great that it has been called the "hallmark" of diving (Murphy et al. 1980). The magnitude of the LA flux qualitatively mirrors the degree of anaerobiosis. To gain a better impression of the meaning of the changes in concentration of LA, and other biochemical changes that occur in diving, and their potential influence on blood flow, a few major features of exercise physiology need to be reviewed.

The following review has relied on several general works in this field (Astrand and Rodahl 1977; Folkow and Neil 1971; Gollnick and Hermansen 1973; Guyton 1981; Newsholme and Start 1973; Hochachka and Somero 1984; Shepherd 1983; Shepherd and Vanhoutte 1975; Skinner 1975). Most of the information in these reviews is derived from man and rats. Information on reptiles and birds is very limited. What is said below is presumed to be true in a general way for these groups as well.

8.1 Muscle Physiology

The following dimensions give an appreciation of circulatory and metabolic requirements of skeletal muscles. About 30 to 50% of the terrestrial and marine mammalian body mass is skeletal muscle. It is approximately 30 and 40% in the adult female and male southern elephant seals, *Mirounga leonina*, respectively (Bryden 1972). Collectively the muscles represent the largest organ of the body, but at rest they consume only 20% of the O_2. In comparison, the liver and kidney consume 40 to 50% of O_2 and in primates the brain requires 10 to 15% of the resting O_2. Blood supply to the resting muscle is about 15% of cardiac output. Yet, in maximum steady state exercise muscle blood flow may be 80% of cardiac output as its blood flow increases four to five times and metabolism increases 10 to 15 times, or 90% of the total O_2 consumption. In bursts of activity muscle metabolism may exceed the blood transport of O_2 by three to five times (Folkow and Neil 1977). Also, different types of muscle place different demands on the fuel supply and storage system.

8.1.1 Morphology

The primary unit within skeletal muscle is the muscle fiber, of which there are three kinds: slow twitch oxidative (ST), fast twitch glycolytic (FT), and fast, oxidative glycolytic (FOG). ST are characterized by slow contractile elements, low myosin ATPase, high oxidative capacity, and low glycerophosphate activity (low glycolytic capacity), high myoglobin content, and low buffer capacity. FT have the opposite characteristics and FOG tend to have features of both; a high oxidative capacity, as well as a high glycolytic capacity, a high myosin ATPase, high myoglobin content, and high buffer capacity.

The muscle bundle, or fasciculus, is made up of a mosaic of these different types of fibers. Athletes may be predisposed to their specialty by the proportion

90

of these fiber types, i.e., sprinters may tend to possess more FT fibers and long distance runners more ST fibers. Harbor seals have a ratio of about 40 ST to 60 FT (P.J. Ponganis, unpubl.) whereas the bottlenose dolphin, *Tursiops truncatus*, has a ratio of about 50 ST to 50 FT (Bello et al. 1985).

8.1.2 Energy Sources and Rates of Utilization

During initial muscle contraction, or periods of intense muscle activity when there is an insufficient supply of O_2, much of the energy is derived from anaerobic glycolysis. There is no acceptable method to determine the amount of contribution from anaerobic and aerobic means, but in maximum work of various durations the ratio is roughly 85 anaerobic/15 aerobic for 10 s, 65/35 for 1 min, 50/50 for 2 min and 10/90 for >10 min (Astrand and Rodahl 1977; Gollnick and Hermansen 1973).

The source of anaerobic energy for the contractile elements in situ ATP is creatine phosphate (CP) and glycolysis to LA. The concentration of CP is about four times that of ATP, and in the course of a short burst of maximum energy output CP may be nearly depleted within 10 to 20 s as it phosphorylates ADP to ATP. In the meantime there is only a small change in ATP. Anaerobic glycolysis may assume the major role of energy production for about 2 min, and at an energy output about half that provided by CP. If the exercise is to go beyond 2 min duration then oxidative metabolism must be the major source of energy. These roles are due to the rates of energy production possible by the different substrates, which if expressed as mmol ATP g^{-1} min^{-1}, are graded from over 100 for hydrolysis of CP and ATP through anaerobic glycolysis (60), aerobic glycolysis (30), to fatty acid oxidation (20); and to the amount of fuel available within the muscle. This ranges from small stores of 10 µmol $\sim P$ $g_{d.w.}^{-1}$ (dry weight) for ATP and 60 µmol $\sim P$ $g_{d.w.}^{-1}$ CP, to large stores of glycogen and tryglyceride, which contains 14,200 and 24,250 mmol $\sim P$ $g_{d.w.}^{-1}$, respectively (Hochachka and Somero 1984).

8.1.3 Lactate

During intense, short episodes of muscular activity when glycolysis is the major source of energy for replenishment of ATP, the FT fibers deplete first, whereas at lower work levels of about 65% of maximum $\dot{M}_{O_2}$, ST are the first to deplete, which requires about 3 h. In the course of these activities at different intensities, the LA produced diffuses across the cell membrane and into the blood, but there is a gradient between the two that can be significant (Hogan and Welch 1986). The blood concentration is dependent upon the rate of production, diffusion, and removal. At 50% of maximum $\dot{M}_{O_2}$ blood LA concentration does not change. From 50 to 85% there is a rapid increase in blood LA, which after a few minutes remains constant or decreases. Above 90% of maximum $\dot{M}_{O_2}$ there is a continuous increase in blood LA concentration. The concentration is also influenced by the amount of muscle involved. Higher concentrations occur when only small groups of muscle result in the same O_2 uptake. After maximal exertion the concentration of LA decreases more rapidly if some exertion continues, and it is greatest, four times that of rest, at about 60 to 70% of maximum $\dot{M}_{O_2}$ (Gollnick and Hermansen 1973).

For some time it was thought that liver was the sole site of glucose production and release (Wahren 1977). However, it has since been shown that FOG and FT fibers of rats can convert LA to glycogen (McLane and Holloszy 1979). Furthermore, in human leg exercise experiments it was shown that the active skeletal muscle was a major site of blood LA removal (Stanley et al. 1986). Also, the total LA release is about two times the net LA release, thus calculations of LA output from the product of blood concentration times blood flow will underestimate total LA production considerably.

Turnover of LA in situ may be due to ST fibers consuming LA generated by FT fibers. Stanley et al. (1986) speculated that during heavy exercise and reduced splanchnic flow the working muscle may remove 60–80% of total LA produced. This view is shared by Brooks (1985), who suggests that more LA than glucose is shuttled during exercise, that release of the former from skeletal muscle is quantitatively as important a carbohydrate fuel source as the release of glucose from the liver, and that a large percent of the LA that is produced is eventually oxidized. The production of LA in exercise is not due necessarily to oxygen lack, but rather to the increase in pyruvate concentration and the high affinity of muscle lactate dehydrogenase (LDH) for pyruvate so that thermodynamic equilibrium favors conversion to LA.

In humans accumulation of blood [LA] is tolerable up to no more than about 18 mM before a subjective feeling of fatigue occurs. If the work is intermittent rather than continuous the tolerance is greater and it is possible for blood [LA] to reach 32 mM (Gollnick and Hermansen 1973). In comparison muscle concentration is about 25 to 30 $mmol\,kg_{w.w.}^{-1}$ (wet weight) after the most vigorous exercise. For perspective it is noteworthy that in the thoroughbred blood and muscle [LA], after four 600-m sprints spaced 5 min apart, was 27 mM and 53 mM respectively (Snow and Harris 1985).

8.1.4 Glucose and Fatty Acids

For maximum, or near maximum burst-type of movement anaerobic glycolysis is suited as an energy source because the energy production rate, as pointed out above, is three times greater than fatty acid oxidation. However, the low storage capacity of the body for glucose and glycogen because it binds with water and occupies a large volume compared to fat makes it an unsuitable source of energy for long-term needs. Also, in carnivores such as marine birds and mammals there is no, or little carbohydrate in the diet, so glucose must be derived from other precursors. Consequently, glucose is probably a premium fuel for special needs, and fat is the all-purpose energy source. For example, carbohydrate provided <5% of the resting metabolic fuel in resting harbor seals (Davis 1983).

For long-term work fat is ideal. It is energy-dense since no water is involved in its storage, and it is readily available for activities with low work rates. The shifts in metabolism during endurance training are exemplary. Such training results in an increased utilization of fatty acids and triglycerides (Holloszy and Booth 1976).

As a final note, fat as fuel may be a liability in aquatic animals during exceptionally long dives in which the O_2 reserves may become critical. It has been shown in myocardial studies of rats and dogs that the O_2 requirement is higher if perfused with high concentrations of free fatty acids. This can be as much as 25% in the dog and 40% in the rat (Vik-Mo and Mjos 1981). Some of this so-called "O_2 wasting effect" is due to the difference in O_2 requirements for complete oxidation of fatty acid and glucose. About 11% more O_2 per carbon unit is consumed by fatty acid oxidation. It has been suggested that this may be significant in the ischemic heart of dog or man. Perhaps it could be an important factor during extended dives.

8.1.5 Control of Blood Flow

The above discussion reveals that the state of training or body condition, rate of energy consumption, the type of organ, as well as the level of blood perfusion, affect the proportion of fuels used. The control of blood flow to skeletal muscle is influenced in three major ways: (1) local, (2) nervous, and (3) humoral. In terrestrial mammals there is extensive, sympathetic innervation of arterioles that control the volume of flow to skeletal muscle. The sphere of sympathetic influence is much greater in diving birds and mammals in which there is a high degree of innervation at the arterial level (Folkow et al. 1966; White et al. 1973; see Chap. 3). The role of sympathetic input and the local milieu have been the subject of numerous reviews, but what the key factors are is still debated (Olsson 1981; Shepherd 1983; Skinner 1975; Sparks and Belloni 1978).

8.1.5.1 Sympathetic Innervation

In resting muscle neurogenic and myogenic factors dominate the control of muscle blood flow. Sympathetic stimulation increases release of norepinephrine that acts on α-receptors in smooth muscle of arterioles, causing contraction and vasoconstriction. In opposition, epinephrine acts on β-receptors, which cause vasodilation. These influences are mediated in numerous ways. For example, baroreceptors, chemoreceptors, and local myogenic responses are factors in the level of resistance in peripheral vessel: (1) If there is an abrupt increase in perfusion pressure there is an immediate, transient increase in blood flow followed by a return to the previous flow. The response is due possibly to an intrinsic smooth muscle reaction to distention of the vascular wall (Skinner, 1975). (2) When arterial P_{CO_2} increases above 5.2 kPa (39 mmHg) or P_{O_2} decreases below 9.3 kPa (70 mmHg) there is an elevation in peripheral vascular resistance (PVR) (Shepherd and Vanhoutte 1975). The degree of these responses is dependent, furthermore, on other local vasodilator factors.

8.1.5.2 Metabolites

The local substances responsible for exercise hyperemia have not been completely resolved (Shepherd 1983; Skinner 1975), but a combination of several metabolites or conditions is implicated. One of these is low O_2 tension. Hypoxia would seem to be a logical, primary candidate, but it fails to meet all the criteria. Nevertheless, hypoxia enhances the dilatory response to reduced arterial pressure and reduces the constrictor response to elevated pressure, but various evidence indicates that O_2 insufficiency does not act directly on vessel muscle (Olsson 1981). The influence may be indirect and due to an increase in vasodilator metabolites as a result of low O_2 availability (Sparks and Belloni 1978).

Numerous other metabolites or factors have been implicated. Some are LA, adenosine, K^+, and increased osmolality. None alone meets all the requirements of a specific vasodilator, which suggests a complex of factors that is further complicated because there may be interspecies variation (Shepherd 1983). An example of the multifactorial effect is provided by evidence that the greatest reduction of vascular resistance in skeletal muscle occurs when the tissue is perfused with hypoxic, hyperkalemic, and hyperosmotic blood. No other combination of one or two of these conditions matches the response to the three together (Skinner 1975).

Such a complex problem and the technically difficult procedures necessary to elucidate control of muscle blood flow seem especially formidable for an understanding of the response in diving animals. Although a considerable effort has been expended on the subject in forced submersion (see Chap. 3), these results may give a narrow view of what occurs in natural dives. The rest of this chapter will show both how little is known, and that blood flow control is far more complex than has been concluded from forced submersion studies.

8.2 Reptiles

Having just stated how much more complex muscle metabolism and blood flow may be in diving animals than in those forcibly submerged, I will give the nearest natural example that I know of a dive response that probably corresponds to a forced submersion. When the land iguana takes refuge in water from an aerial or terrestrial threat it may remain submerged for up to 4.5 h. In that time span it has been observed that blood [LA] rises to 35.2 mM from about 3 mM (Moberly 1968). Such a high value was unusual. Most voluntary submergences resulted in a blood concentration rise of < 25 mM (Fig. 8.2).

Unlike a terrestrial animal that is using the water as a place to hide, aquatic vertebrates use it as a place to extract a living. The result is a different behavioral and metabolic character of diving. In the marine iguana the animals dive for relatively short periods of 5 to 10 min. During much of that time they cling to bottom rocks while they graze on algae. Blood [LA] obtained from lizards returning from foraging dives differed little from resting levels (Gleeson 1980). This was some-

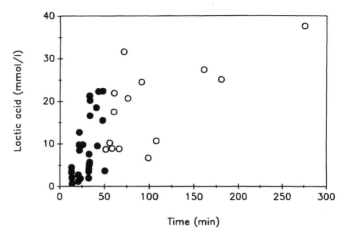

Fig. 8.2. Blood lactate concentrations in relation to submersion duration in the common iguana, *Iguana iguana. Open circles* are values from lizards that remained submerged until "exhausted." These lowest values of both open and closed circles are from animals that were inactive while submerged. (After Moberley 1968)

what true for the sea snakes, *Laticauda laticauda* and *L. colubrina*, which were also captured just after dives of unknown duration. Only occasionally, in two out of twelve samples, were the [LA] of 12.2 mM notably higher than the others of 3.8 to 6 mM. No resting measurements were made, but it was presumed they would be <1.7 mM. The high [LA] requires about 4 h to return to normal (Seymour 1979).

The recovery duration is noteworthy because during tracking of yellow-bellied sea snakes it was found that 21% of all dives (202) exceeded an hour in duration, and one lasted 213 min (Rubinoff et al. 1986). Some of these dives exceeded the known survival time of forced submersions, and a large proportion (75%) exceeded the calculated aerobic dive limit. However, these snakes dived repeatedly, and did not require long recoveries at the surface, even though some serial dives were each more than 1 h in duration (see Fig. 13.1). Determination of blood [LA] after these dive patterns would be interesting, especially in view of the LA recovery pattern of other sea snakes as well as the sea turtle.

Recovery may be even longer in sea turtles. After forced submersion peak arterial [LA] of about 7.8 mM required over 15 h to return to pre-dive levels in the green sea turtle (Berkson 1966; Fig. 8.3). However, circulation in the sea turtle after forced submersions tends to be sluggish, which is indicated by failure of [LA] to reach maximum until 1 h after the dive. If slow recoveries from LA accumulation seem to be distinctive for reptiles and sea turtles in general, the singular attribute of the freshwater turtle's prolonged submersion is the tolerance to LA.

Normally the dives of freshwater turtles of various species may not be long. At summertime temperatures of 22 to 25 °C, most dives of *P. scripta* were <10 min and not more than 44 min. During this time the alveolar P_{O_2} was always greater than or equal to 22 mmHg, which indicated an oxygen tension threshold, and also that the dives were aerobic (Ackerman and White 1979). The

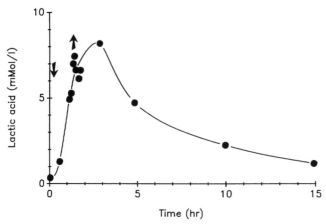

Fig. 8.3. Blood lactate concentrations during about a 1-h forced submersion in a green sea turtle, *Chelonia mydas. Arrows* indicate the start and end of submersion. (After Berkson 1966)

aerobicity of dives is supported by whole-body [LA] measurements of *C. picta* which, during summer dives at 25 °C, averaged 2.2 mM (Gatten 1981). When animals took refuge underwater because of aerial threat, the whole body LA rose to 8.8 mM by 60 min, a value not too different from other divers during prolonged breath-holds.

The above results all seem conventional for vertebrate divers, but on the other hand long-term submergence, anoxic tolerance (Chap. 2) and LA tolerance are phenomenal. During wintertime torpor, in which the turtles were underwater for 67 days at 3.7 °C whole body, [LA] rises to at least 62 mM (Gatten 1981). Under similar conditions, with the same species, but for 12 weeks, plasma [LA] has reached 145 mM (Jackson and Heisler 1982). At this level the HCO_3 available cannot balance this ion shift and much comes from Ca^{2+} and probably Mg^{2+}. These spectacular ion shifts and metabolite concentrations are of much interest .to acid-base physiology and ion balance, but they tell us little about diving responses and flow distribution, the principal objective of this chapter.

The iguana hiding underwater, or the turtle resting for short, aerobic periods below the surface may be examples of two types of response at opposite ends of a spectrum. The iguana probably has greatly restricted blood flow and reduced metabolism, whereas the turtle resting in summertime, or the sea snakes diving may have quite a different blood flow pattern and a metabolism that is not reduced, but may even have been elevated. There is only one report of measurements of muscle blood flow during diving or submersion in a reptile, or in any other type of diver as far as I am aware.

Muscle blood flow was determined in the tail of 3- to 4-kg American alligators, *A. mississippiensis*, held in small tanks at 25 to 35 °C. By means of xenon clearance it was shown that blood flow during forced submersions of about 7 min was a third of flow during resting submersion in which the latter lasted no more than 3 min (Weinheimer et al. 1982). It was concluded that cessation of peripheral (muscle) blood flow is a response not to diving, but a reaction to fear and escape.

96

8.3 Birds

There is little information on blood or muscle metabolites, or muscle blood flow distribution during diving in birds. There is some irony therein because one of the first challenges to the concept that forced submersion blood flow and metabolite variations are representative of natural dives was due to studies on birds. During his thesis study of several species of diving birds which included shags, guillemots, eiders and razorbills, Eliassen (1960) obtained results that caused him to question the accepted view of depletion of blood and muscle O_2 store, decrease in metabolic rate and accumulation of blood and muscle LA. Some of his evidence was that both blood and muscle [LA] increased more or less in parallel, with muscle having a concentration that was always about two times that of the blood during forced submersion. Since then there has been little evidence for or against these proposals in birds except for data on heart rates (Chap. 6). However, as you will read below, blood metabolite data in diving seals support some aspects of his contentions.

8.4 Mammals

The most detailed information on intermediary metabolite levels and changes, especially of LA, come from investigations on mammals. Much is from forced submersion experiments, which help to interpret the few works on natural dives. All results, except for one experiment, are blood values. In a report discussed previously (Chap. 3) it was shown that muscle LA accumulation and O_2 depletion curves tend to mirror each other (Scholander et al. 1942). As muscle O_2 declines, anaerobic glycolysis increases. Also, the longer the submersion, usually the higher the peak of arterial [LA] after the submersion. These blood LA features after a dive or forced submersion are a historical indicator of the relative anaerobic contribution during the breath hold, and because of this it will receive much emphasis in the following discussion, as it has in previous parts of this chapter. To give some perspective to comments on natural diving, a few remarks on specific sites of metabolism should be helpful.

8.4.1 Brain

Calculations based on body and brain size will show that the brain of the Weddell seal accounts for only about 1% of the total energy metabolism compared to 15% in man (Hochachka and Somero 1984; Kooyman 1981). As an aside, almost all of the rest of this chapter will center around the Weddell seal because it has been studied in such detail.

The fuel of the mammalian brain is almost entirely glucose, 75% of which is metabolized oxidatively. The other 25% is a constant anaerobic input, which in the seal does not change until the arterial P_{O_2} falls well below 3.3 kPa

(25 mmHg), or slightly less than half saturation. Indeed, increased LA production was not apparent until after 19 min of forced submersion in a harbor seal whose breath – hold limit was 23 min. The breath-hold limit was based on an endpoint criteria in which cerebral brain waves became anomalous (Kerem and Elsner 1973). Similar endpoints determined for the Weddell seal occurred after forced submersions of 42 and 54 min, and an arterial P_{O_2} as low as 0.9 kPa (7 mmHg) (Elsner et al. 1970 b, Chap. 2). The arterial saturation at this point would be $<10\%$ (Lenfant et al. 1969 b). This limit, by forced submersion and chemical restraint, is less than the longest dive measured of 73 min (Kooyman et al. 1980). Even in such long-breath holds, normal low O_2 and glucose consumption, and LA production of the brain would not significantly alter the blood levels of these metabolites (Hochachka and Somero 1984; Kooyman 1981; Kooyman 1985).

8.4.2 Heart

Another possible contributor of LA to the central circulation is the heart. According to some investigators, the Weddell seal heart relies solely on oxidative pathways except towards the end of the most extreme submersions (Murphy et al. 1980; Hochachka and Somero 1984). However, LA production by the heart of harbor seals occurs throughout the forced submersion, based on concentration differences between the aorta and coronary sinus (Kjekshus et al. 1982). The anaerobic potential may be indicated by the fact that the heart has one of the highest concentrations of stored glycogen in mammals. Cardiac muscle of the adult Weddell seal contains 22 mg glucose $g_{w.w.}^{-1}$ compared to several domestic mammals which range from 2.4 to 6.8 mg glucose $g_{w.w.}^{-1}$ (Kerem et al. 1973). Studies by Kjekshus showed that LA production by the heart is not great relative to total body production during forced submersions. During recovery, when LA concentrations are exceptionally high, the heart becomes an avid consumer of LA, based on concentration differences. The consumption rate may be ten times the pre-submersion level (Kjekshus et al. 1982).

8.4.3 Lung

Another consumer of LA, but in this case during the submersion as well, is the lung, which appears to oxidize LA at a rate of 0.5 mM min^{-1} (Hochachka 1980). The lung, therefore, might be responsible for a check on accumulation of LA in the central blood pool during submersion, but there is no evidence for glucose synthesis (Murphy et al. 1980) as originally suggested (Hochachka et al. 1977).

8.4.4 Glucose

Metabolic rates of three organs, brain, heart, and lung account for the consumption of only a small proportion of the O_2 store during a dive, which otherwise could, if dependent on only these organs, last nearly 3 h (Kooyman 1981; Chap. 12). Further, these organs can account for only a small part of the increase in LA and decline in blood glucose during a submersion. Although blood glucose consistently declines in forced submersions (Murphy et al. 1980; Kjekshus et al. 1982), it is less so in natural dives. Mean whole blood glucose values of resting Weddell seals were 4.0 mM in adults and 5.1 mM in sub-adults (Castellini, Davis and Kooyman 1988) and 6.2 mM in plasma of adults (Guppy et al. 1986). In adults 2 min after surfacing from dives ranging from 10 to 61 min the average value was 4.1 mM (N = 15), and 4.5 mM (N = 132) in 4- to 30-min dives of sub-adults (Castellini, Davis and Kooyman in press), and 2.5 to 4.2 in dives ranging from 8 to 30 min of adults (Guppy et al. 1986). The difference between the two groups of investigators is due in part to a measurement of whole blood glucose and plasma. Red blood cells of seals have low glucose content and are also a large part of whole blood volume. The result is that a measurement of whole blood glucose will be approximately 50% of a plasma measure (Hochachka et al. 1979).

There is a steady increase in blood glucose after extended dives (Kooyman et al. 1980), or forced submersion (Davis 1983; Hochachka et al. 1977; Kjekshus et al. 1982; Robin et al. 1981). This increase is possibly due to an outpouring of glucagon and catecholamines (Robin et al. 1981); it occurs immediately after the dive with one exception. In some experiments with Weddell seals glucose continued to decline for about 10 min post-submersion (Murphy et al. 1980). The reason for the difference in these results is not clear, but important because of interpretation.

A continued decline in glucose concentration was interpreted as evidence that the total blood volume was depleted because of large consumption of glucose in tissue other than heart, lung, and brain. Further, it had to be anaerobic because the consumption rate was so high that it had to be explained by inefficient substrate usage (Murphy et al. 1980). However, the reason for the continued fall of blood glucose after the submersion is not explained. At this time tissues are once again well oxygenated, and a high blood LA concentration promotes its utilization as a favored substrate by some organs (Kjekshus et al. 1982; Hochachka 1980). Also, the post-submersion levels of glucose in the experiments discussed above fell to as low as 2.2 mM (a life-threatening level in other mammals) after only 10 to 20 min of breath-hold, but in voluntary dives of the same species, glucose was never lower than 3.3 mM whether after a single dive of 60 min (Kooyman et al. 1980), or a series of hunting dives (Castellini, Davis and Kooyman 1988). The divergence of these results by various investigators carries a different emphasis on the importance of glucose as a fuel, and how it is combusted.

The opinion was expressed that after dives of about 5 min or greater in the seal, blood glucose is the "major, if not the only, source of carbon for metabolism" (Hochachka 1980). This is perhaps a logical conclusion from what was formerly a widely shared judgment, i.e., that although diving involves a mixture of aerobic and anaerobic metabolism, it is mainly anaerobic, and that the salient features of diving are: (1) LA accumulation in muscle during the dive, and (2) its washout and abrupt increase and slow decline in blood after the dive (Hochachka and Murphy 1979). The statement is accurate for forced submersion in almost all reptiles, birds, and mammals studied, but it is not true for voluntary dives.

8.4.5 Lactate

I alluded earlier to the usual absence of LA after dives of unknown duration in sea snakes and marine iguanas. More certain and striking evidence comes from the difference between forced submersions and natural dives of Weddell seals.

Using procedures for the investigation of the physiology of voluntary diving described in Chapter 6, arterial blood samples were obtained from adult and immature Weddell seals ranging in size from 130 to 450 kg. LA endurance curves were constructed by plotting maximal peak post-dive arterial [LA] against duration of dive (Fig. 8.4). For dives up to 5 min for yearlings (130 to 145 kg), and to about 20 min for adults (350–450 kg) there was no correlation between dive duration and LA concentration because the concentration was not different from resting values (Guppy et al. 1986; Kooyman et al. 1980; Kooyman et al. 1983). Dives in excess of those durations resulted in a good correlation with the LA concentration (Kooyman et al. 1980; Kooyman et al. 1983). This was the first time, and the only species, in which such measurements have been obtained due to the special conditions required to conduct such experiments (Chap. 6). It was also the only determination, by a physiological variable, of the aerobic dive limit (ADL). The curves also indicate a size relationship which is discussed more in Chapter 10. It is interesting that results obtained for adults by another group (Guppy et al. 1986)

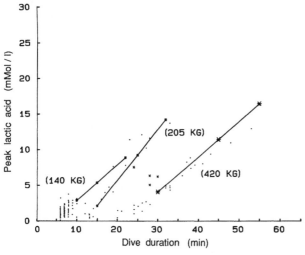

Fig. 8.4. The relationship of dive duration and maximum post-dive blood lactate concentration in adult and immature Weddell seals. At the point of departure from resting to the highest values the regression equations in order of the smallest (140 kg) to the largest (450 kg) Weddell seals are: $Y = 2.1 + 0.5 X$, $Y = -8.5 + 0.71 X$, and $Y = -10.7 + 0.49 X$, (data from Kooyman et al. 1980; Kooyman et al. 1983). The *asterisks* are values from another group that studied 300 to 400 kg Weddell seals in a similar manner. (Guppy et al. 1986)

tend to be above the adult curve. Possibly the samples are from a seal < 350 kg, and/or the added drag from the sampling pack caused more work. These results of blood [LA] compel a re-evaluation of intermediary metabolism during the dive and the types of fuel. This point is discussed further in the summary and conclusions, after a few other aspects of the post-dive recovery are reviewed.

Due to an unusual and fortunate coincidence when physiological variables of diving were being investigated in Weddell seals, another group was measuring blood metabolite change during forced submersion. This made possible some interesting comparisons. The LA recovery curve of both a 46-min forced submersion and 43-min dive show that the recovery is quicker, and the peak lower after the dive (Fig. 8.5). The more rapid recovery is due to anticipated surfacing and increased blood flow distribution. The lower peak is more difficult to explain, but suggests that: (1) the dive was anticipated and O_2 stores were fully loaded, (2) the slow, methodical swim mode of the prolonged dive was less energetic than the uncoordinated struggling of a restrained animal, and (3) the general condition of the free-diving seal was probably better than the restrained animal which had undergone recent major surgery. These differences in prolonged dives are less perplexing than differences in the shorter dives.

Forced submersions of 10 to 20 min, which are within the ADL of the Weddell seal, result in a distinct blood LA pulse in which the maximum concentration may be 9 mM after a 10-min submersion. Also, there was no relationship between duration and maximum LA concentration. The other four submersions were of longer duration, up to 20 min, but none resulted in as high a post-submersion blood [LA] (Murphy et al. 1980). These results indicate a substantive difference between dives and forced submersions. Since the reasons for these differences are unknown, any explanation would be speculative. However, there is one similarity between forced submersions and dives that was commented on earlier.

In general, after prolonged dives the "classic" post-dive abrupt rise in blood [LA] is followed by a steady decline (Fig. 8.5). This similarity, coupled with the observation that a marked decrease in heart rate occurs immediately after the beginning of a prolonged dive (Chap. 6), demonstrates that the cardiovascular responses may be somewhat the same, i.e., a reduced cardiac output, restricted blood flow to the periphery, and greater reliance on anaerobic metabolism (see Chap. 3). Dissimilar is the post-dive rate of decline in LA. Note that even after

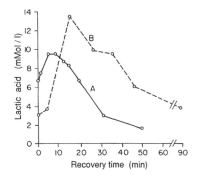

Fig. 8.5. Comparison of changes in arterial lactic acid concentrations during recovery from a 43-min dive (*A*), and a forced submersion of 46 min (*B*) in adult Weddell seals. (After Kooyman 1982)

90 min the forced submersion value is high and rate of decline after 50 min of recovery becomes different from the dive recovery.

Certain variations in the decline of LA after a prolonged dive also provide evidence for the way in which dive response may work in shorter dives. If it is assumed that this curve is a normal, undisturbed recovery rate, then a comparison with two other curves is provocative. Soon after surfacing from a 53-min dive, an adult seal dived for 10 min. Blood LA continued to fall during the dive, but rose again after the dive to nearly the concentration before the 10-min dive concentration (Fig. 8.6). Apparently blood flow to muscle, the source of the blood LA, was restricted during the 10-min dive. Since Weddell seals seldom make a dive again so soon after such a long dive, it may have been a startled reaction to another seal, or the investigators drawing blood samples, that resulted in a moderate to strong dive response. In contrast to this decline and increase in blood LA after a dive, which interrupted recovery from a prolonged dive, there was a steady decline in blood LA after a 33-min dive of a sub-adult that continued to dive so regularly that it suggested foraging dives (Fig. 8.7). The maximum depth of the first dive was 220 m, and the maximum depth of the others was 190 m. This same type of recovery occurred after a dive of 25 min. When these two curves and that for the animal that made only one 11-min dive soon after surfacing from a 28-min dive are compared, it can be seen that all had about the same rate of decline in blood [LA] (Fig. 8.7).

The rate of clearance in the diving and swimming animal was no more rapid than in an animal that remained at the surface for the entire recovery. This may indicate that during serial dives the recovery can be described as one of constant high rates of ventilation and perfusion that results possibly in an overall ventilation and cardiac output equal to that of the seal that remains at the surface for the whole recovery. Thus, a seal can and will continue to dive with a LA burden, but if the subsequent dives remain within the seal's aerobic dive limit the LA will continue to decline. In this particular seal, as soon as the blood [LA] after the 25-min dive returned to a resting concentration it made another long (28-min) dive.

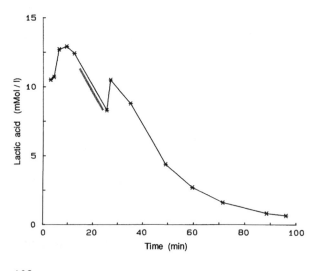

Fig. 8.6. Arterial lactate concentration during recovery from a 53-min dive by a Weddell seal. The *hatched bar* indicates the interruption of the recovery by a 10-min dive. (After Kooyman et al. 1980)

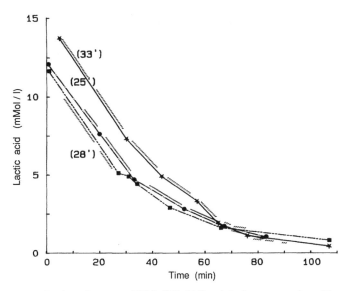

Fig. 8.7. Arterial lactate concentration in an immature (200-kg) Weddell seal during recovery from 25-, 28- and 33- min dives while short, hunting dives continued. The *hatched segments* represent the dive times. (Data from Castellini et al 1988)

Fig. 8.8. Hypothetical blood lactate concentration during a series of dives in an adult, female northern elephant seal which exceeded its presumed aerobic dive limit. The *numbers near the slopes* are the dive durations and the *interval* between the low and high points of the slopes is the surface interval (about 2.2 min). (Modified dive data from LeBoeuf et al. 1988)

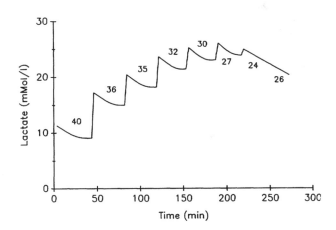

Based on behavioral information, a diver will not continue to exceed its aerobic dive limit (ADL) after an initial extended dive. An exception may be one case in the northern elephant seal (Le Boeuf et al. 1988). Because the female was equivalent in size to a Weddell seal (400 kg), I assume it had about the same ADL. If so, it suggests that blood [LA] may have risen to high levels before the bout of extended dives was over. It began with a 46-min dive, immediately followed by five more dives beyond the estimated ADL in which the surface ventilation between each dive was 2 to 3 min. Based on decline of [LA] in the Weddell seal after an extended dive (Figs. 8.6 and 8.7) I calculate a hypothetical case in which blood [LA] may have risen as high as 26 mM in the course of 3 h of diving

103

Table 8.1. Estimated total muscle LA production during a 65-min dive of a 450 kg Weddell seal. Also estimated is the maximum blood [LA] during recovery and after equilibration of blood and tissue water. These calculations consider only production not turnover and consumption of LA. V_b = blood volume, V_m and V_t = muscle and tissue water volume

---

Muscle aerobic energy requirement (mM O_2 kg^{-1} min^{-1})[a]
0.11 (70% muscle mass at rest) + 0.22 (30% at 2 × rest) = 0.14

$\dot{M}_{ATP}$ = 0.84 mM ATP kg^{-1} min^{-1} [b]
Anaerobic equivalent
$\dot{M}_{LA}$ = 0.84 mM LA kg^{-1} min^{-1} [c]

50 min anaerobic duration (total 65 min–15 min aerobic)
Tissue concentration of LA
Muscle [LA] = 0.84 × 50 = 42 mM LA kg^{-1} (wet weight), or

42 mM LA kg^{-1} ÷ 0.73 H_2O fraction = 58 mM LA l^{-1}
Tissue [LA] = 0.66 mM LA kg^{-1} min^{-1} (w.w.), or
33 mM LA kg^{-1} ÷ 0.73 H_2O = 45 mM LA l^{-1}

Muscle and tissue water volume = 160 l
(100 l muscle + 60 l other organs)

Blood volume = 70 l
Post-dive equilibrium
      Blood   Muscle   Tissue
(70 l · 1 mM LA l^{-1}) + (100 l · 58 mM LA l^{-1}) + (60 l · 45 mM LA l^{-1})
  ÷ (V_b + V_m + V_t) = 37 mM · LA · l^{-1}

---

[a] Kooyman et al. (1973 a); Kooyman (1981).
[b] 6 mol ATP per 1 mol O_2 consumed.
[c] 1 mol ATP per 1 mol LA produced.

(Fig. 8.8). It then would have declined steadily to normal levels over the next 2 h in which continuous aerobic dives prevailed.

Finally in an attempt to determine just how high muscle [LA] might become at the end of the most extreme dives, calculations were made of blood and muscle concentration for a 65-min dive (Kooyman 1985). Muscle [LA] would be 42 mM kg$_{w.w.}^{-1}$ (Table 8.1). This is substantially lower than that mentioned earlier for sprinting horses (Snow and Harris 1985). It also does not account for any turnover of LA and its consumption by both muscle and other organs as occurs under other conditions (Brooks 1985; Stanley et al. 1986). Omitting any turnover considerations would give a worst possible case, and indeed maximum blood [LA] calculated of 37 mM (Table 8.1) is higher than that of 26 mM obtained in a seal after a 61-min dive.

8.5 Conclusions

At present there is no information on changes in concentrations of metabolites in blood and muscle during diving in birds, and little in reptiles. The information available for mammals comes from studies on seals. Although this may seem narrow, many of the features of diving are probably universal for most divers and what is said below may have general applications.

First, the evidence presented in previous chapters, and now in this chapter, supports the hypothesis that the dive response is not a simple on-off reflex, of, for example, restricted blood flow and aerobic to anaerobic metabolism. It is a complex of neural and metabolic activities that are greatly modified by circumstances and the objectives of the diver. This complexity raises many more questions about management of O_2 stores and fuels than the previous concept of diving derived from forced submersion experiments only. In some respects it may seem a setback that much less is known about diving physiology than was previously thought, but in another way it reveals many new and exciting prospects to learn about the way in which animals dive. The following is an attempt to suggest some of the ways by which divers function based primarily on results presented in this chapter.

In burst activity diving animals probably exhibit the normal sequence of biochemical processes that supply energy to muscles. There is a high energy output over a period of <1 min in which hydrolysis of CP and ATP are the major source of energy for muscle contraction. If the work effort persists for several minutes, then anaerobic glycolysis assumes the predominant role. For long-term efforts of minutes to hours muscle contraction is supported by a lower rate of energy output and metabolism is oxidative. The fuels grade from CP to glycogen to triglycerides and free fatty acids (FFA), and in the common vernacular of track and field events, from the sprint to the marathon.

The complementary water activities would be the short, burst swim, to the long migratory episodes in which surfacing is a rhythmic activity to meet ventilatory needs. This is the surface condition, but what is the situation during diving? The extended breath-hold is primarily a sprint in slow motion, but it has attributes of the marathon as well. Different from a sprint, the swim rate is low, like a marathon, and in slow motion muscle energy is released. The event may last more than an hour in a diver like the Weddell seal, which may seem long like a marathon, except that the normal dive bout is 10 to 15 times longer. The fuels during the extended dive, and the recovery after are similar to a sprint. An important source of fuel is supplied by anaerobic glycolysis, and there is an extended period after the dive to clear LA, the endproduct of anaerobic glycolysis. The concentration reaches a peak comparable to a heavy work load, and decays at about the same rate (Fig. 8.9)

Distinct from this pattern is the foraging bout in which the seal probably swims at higher rates than during a prolonged dive, but not above a level that muscle contractions cannot be supported aerobically, mainly by oxidizing fat, as does the long-distance runner, thus conserving blood glucose, and muscle and liver glycogen stores. The dives are of relatively short duration within the ADL, and surface recovery is brief and used primarily for loading O_2 stores. Unlike the marathon runner, who paces himself to reach the end of the event as muscle glycogen stores deplete, the bout ends due to satiation from a successful hunt. Indeed, in some the bout may never end, as animals such as northern elephant seal and leatherback sea turtles seem to dive almost continuously (see Chap. 13). The

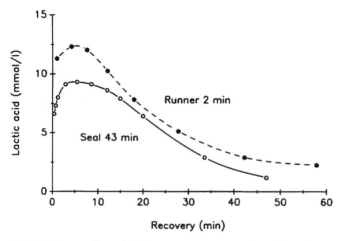

Fig. 8.9. A comparison of the lactate recovery curves of a 2-min sprint by a human athlete and a 43-min extended dive by a Weddell seal. (Data from Astrand and Rodahl 1977; Kooyman et al. 1980)

fuel supplies in these different swim events of the diver are subjects of speculation.

How are these fuels mobilized and distributed? Serial, short duration dives suggest the greatest possible differences from forced submersions, or extended dives, and add a level of complexity referred to earlier. Heart rates reviewed in Chapter 6 indicate that there is some reduction in cardiac output during aerobic dives, but splanchnic flow appears to remain near normal levels (Chap. 7). Is the reduction in cardiac output due to a reduced perfusion of the muscles?

This question, crucial to understanding the response to natural diving, remains unanswered, but I would like to propose a model that considers local muscle O_2 and fuel stores, without compromising access to O_2 and metabolites external to the muscle. A key reason for the model is to resolve the question of why the concentrations of myoglobin (Mb) in diving animals (Table 5.1) are so high, unless it is to function as an O_2 store. Indeed, a perusal of Table 5.1 shows that an increase in Mb concentration is more consistent in divers than an increase in hemoglobin. For Mb to participate in this role, flow must be pulsatile, so that O_2 is stripped from the Mb molecule during the no-flow condition. If it were otherwise and flow were continuous, Mb would remain saturated because of its much greater affinity for O_2 than Hb; $P_{50} = 0.7$ and 4.0 kPa, respectively (Fig. 9.1). This type of O_2 utilization is different from terrestrial mammals only in magnitude of the no-flow condition, which lasts only during the contraction phase in a terrestrial animal. When moving, this is brief, but when contraction is tonic to support a load, contraction can be extended and the muscle may fatigue sooner.

The model proposed follows general mammalian features except that restriction of blood flow is at different levels (Fig. 8.10). As the diver submerges, flow in skeletal muscle ceases due to vasoconstriction of arterioles as described earlier in the general mammalian circulatory control. Duration of flow restriction is dependent on the work level of the muscle, and the concentration and saturation

106

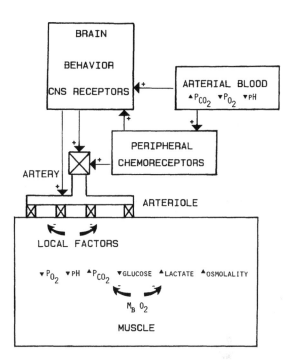

Fig. 8.10. Hypothetical control of blood flow and O_2 delivery to muscle during aerobic hunting dives, and extended exploratory dives. See text for detailed discussion. (Kooyman 1987 b)

of Mb. During most of this time, metabolism is oxidative and the fuel is triglycerides and fatty acids. As muscle O_2 declines glycogen consumption, anaerobic glycolysis, and LA concentration increase. Such local factors cause the release from vasoconstriction and the resulting increase in blood flow facilitates the exchange of metabolites and gases. Possible factors, as in other mammals, might be a decline in P_{O_2} and pH, and increase in P_{CO_2}, [LA],[K$^+$] and osmolality. This accounts for a means of utilizing the O_2 store within the muscle, which in a diver such as the Weddell seal may be sufficient to maintain resting muscle oxidatively up to 15 or 20 min. But what happens when the seal plans, or by accident makes, an extended dive?

When some factor, perhaps arterial P_{O_2}, and/or high CNS regulation, and other augmenting conditions reach a certain level of input, the threshold for arterial vasoconstriction is exceeded and blood flow to muscle and other organs is greatly restricted, proximal to the influence of local factors that act on arteriole myogenic tone. There is little evidence for such regulatory control, but the structures are there: broad vasoconstrictive capacity, large tissue O_2 stores, tolerance to low P_{O_2} and high P_{CO_2} and LA, and high tolerance in some organs to anoxia. Furthermore, the model exposes areas of possible research. Indeed, Castellini et al. (1985) argue that there is a physiological "conflict" between diving and exercise. The needs of diving require restricted flow to protect the central blood O_2 and glucose stores for obligate aerobic and carbohydrate consuming organs, while exercise requires an increased circulation to meet the needs of the increased energy demands of skeletal muscle. Consequently, in a comparison of forced submersion and exercise, Castellini et al. (1985) found a marked difference in the turnover rate of FFA and glucose. If this method were applied to a diving animal it might provide qualitative evidence on if and when peripheral vasoconstriction occurs. This is somewhat akin to evidence presented earlier about the rate and profile of [LA] recovery after an extended dive followed by total rest, or more short dives (Figs. 8.6 and 8.7).

8.6 Summary

There is a message in the fact that this chapter on muscle metabolites and blood flow is perhaps the area about which we know least, but on which, I have nevertheless written the most. Because diving physiology is at an important milestone, I have tried to show through a brief general review of muscle metabolite fluxes and blood flow control that in general, adaptations in amphibious divers may be mainly quantitative differences from terrestrial animals. So far, however, there is little on which to base this statement. Little is known about control of blood flow in skeletal muscle and utilization of O_2 in reptilian divers, nothing on birds, and most of the evidence on mammals comes from one species, the Weddell seal. The results from the Weddell seal compel us to reappraise the dive response in this new light that for most dives there are no major changes in blood chemistry with regard to the glycolytic metabolites of glucose or LA. This suggests that there is no broad vasoconstriction during diving, or there is restriction, but it is controlled in such a way that muscle groups remain within aerobic limits. To explain this a model was proposed, using an extension of basic physiological responses to exercise that account for the aerobic dive limit, the high concentration of Mb in divers, and the absence of a surge in [LA] in blood after dives. This was an attempt to expose one of the most exciting aspects of the biology of natural divers – how they manage their O_2 store.

Chapter 9

Blood Gases

Blood gases are a difficult variable to measure. Unlike heart rate, for example, which can be obtained with a simple electrode contact to the skin, the animal must be catheterized to measure blood gases. This is always a complex procedure in aquatic animals, and sampling while free-diving adds further great complications. Recently, blood gases have been obtained during dives of Weddell seals. This was done by means of a microprocessor-controlled sampling array. It is a major innovation in the study of the physiology of diving. Some of the LA and glucose results were presented in Chapter 8; the blood gas results are discussed below.

One nonintrusive means of indirectly obtaining the end of dive P_{O_2} and P_{CO_2} is to measure end-tidal gas samples, and assume a small difference between alveolar and arterial gas tensions. The sampling, which requires special conditions, has been done in only two species of marine mammal; the Weddell seal and bottle-nosed dolphin.

Furthermore, a validation of the arterial/alveolar difference (Δa-A) has never been done.

All of the inherent difficulties with blood gas measurements mean that there are few data for this important area of diving physiology. Much of what I will discuss relies on data obtained from forced submersions or resting apneusis, most of which are results from seals. Because basic general information on these variables is also mainly from mammals, primarily the human, the following discussion of general properties reflects this emphasis.

9.1 General Properties

Mammalian blood gas properties, which have been extensively studied, are the standard of comparison to the lesser known characteristics of reptiles and birds. Some major features of reptilian blood gases have been reviewed by Wood and Lenfant (1976). Compared to mammals there is little known, and perhaps more important, there is little constancy of levels of blood gas tensions among different groups because of the wide variation in intracardiac shunting. This shunt results in large differences in pulmonary and arterial O_2 and CO_2 tensions. Other sources of variation are the uneven distribution of gases in the lung, the arrythmic respiration, and the range of preferred temperatures of reptiles. Conditions are more consistent in the endothermic birds and mammals.

Blood gas conditions in birds are not known in as great a detail as in mammals. However, they can be presumed to be similar to those of mammals (Scheid

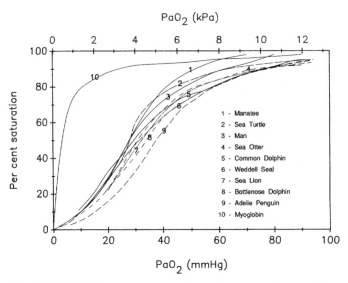

Fig. 9.1. Blood O_2 dissociation curves of the major groups of marine divers and the dissociation curve for human myoglobin. Determined at 37 °C and pH 7.4 for the mammals, 38.5 °C and pH 7.4 for the Adelie penguin and 25 °C and pH 7.47 for the sea turtle. (Data from Bunn et al 1977; Horvath et al. 1968; Lapennas and Lutz 1982; Lenfant et al. 1969 a,b; Lenfant et al. 1970; White et al. 1976)

1979). The normal, resting alveolar O_2 pressure (PA_{O_2}) in the human, at sea level, is 13.9 kPa (104 mmHg). This reflects a decline from the ambient tension of 21.2 kPa (159 mmHg), and it is due to saturation of the inspired air with water vapor, as well as the mixing of lung gases. On exercising, the alveloar ventilation must increase approximately by the same amount as the oxygen consumption to keep PA_{O_2} = 13.9 kPa.

The returning mixed venous blood $(P\bar{v}_{O_2})$ = 5.3 kPa. During passage through the pulmonary capillaries it approximates alveolar tension, but due to a shunt of bronchial circulation of 1 to 2% of cardiac output, the mixed arterial blood equals 12.7 kPa (Guyton 1981). When Pa_{O_2} = 12.7 kPa the blood is about 95 to 97% saturated if pH = 7.4, P_{CO_2} = 5.3 kPa, and temperature = 37 °C (Fig. 9.1). The half saturation tension (P_{50}) equals 3.6 kPa. The interstitial P_{O_2} is about 5.3 kPa, which matches the venous O_2 pressures. At this tension venous saturation is about 70% (Fig. 9.1) and the arterial-venous difference $(\Delta a-\bar{v})$ is about 5 vol.%. This pressure and saturation are dependent upon oxygen consumption and blood flow.

Heavy exercise causes venous tension and saturation to fall to 2.7 kPa and a $\Delta a-\bar{v}$ of about 15 vol.%. Under the varying conditions of rest, exercise, and local environment the dissociation properties of oxy-Hb shift. Factors with the most powerful influence are P_{CO_2}, pH, and temperature. These influences are usually referred to by the eponym "Bohr effect". A large Bohr effect would mean that the dissociation curve shifts to the right markedly with small increases in P_{CO_2} and temperature, and decrease in pH.

110

9.2 Reptiles

As one would expect from the earlier comment on intra-cardiac shunting, the resting blood gas values of reptiles are highly variable. Just before a dive, in the nile monitor lizard, *V. niloticus*, arterial gas tensions were $Pa_{O_2} = 16.0$ kPa and $Pa_{CO_2} = 2.7$ kPa at 30 °C (Wood and Johansen 1974). Blood and lung gas tensions of a freshwater turtle, *P. scripta*, resting at 20 °C, were $Pa_{O_2} = 8.0$ to 9.3 kPa, $Pa_{CO_2} = 3.3$ kPa, $PA_{O_2} = 16.0$ kPa and $PA_{CO_2} = 2.7$ kPa (Burggren and Shelton 1979). Those of adult (87 kg) supine green sea turtle, *C. mydas*, "resting" at 25 °C were $Pa_{O_2} = 6.2$ kPa and $Pa_{CO_2} = 4.0$ kPa (Wood et al. 1984).

9.2.1 Hemoglobin Affinity for O_2

At $Pa_{O_2} = 6.3$ kPa arterial blood in the green sea turtle is 90% saturated due to its high affinity for hemoglobin. $P_{50} = 2.4$ kPa (Fig. 9.1). There may be a size relationship because in 1.1-kg green sea turtles $P_{50} = 4.2$ kPa (Wood et al. 1984), and in 3.4-kg turtles $P_{50} = 3.2$ kPa (Lapennas and Lutz 1982). In contrast to *C. mydas* the P_{50} of the loggerhead sea turtle, *Caretta caretta*, is 6.3 kPa for about a 4-kg animal (Lapennas and Lutz 1982). Thus, P_{50} in the sea turtle ranges from the lowest to the highest compared to nine other species of turtle in which the P_{50}'s ranged from 2.8 to 4.6 kPa at 25 °C and pH $= 7.6$ (Wood et al. 1984). The reason for these wide differences is uncertain.

In an attempt to explain the difference between the green and loggerhead turtles, Lapennas and Lutz (1982) suggested that divers might be divided into two groups; those that depend upon lung O_2 stores and those that rely on blood O_2 store. The high affinity and a low Bohr effect serve well the extraction of O_2 from the lung store at low O_2 tensions. According to the authors, the low affinity of loggerhead blood is compensated by a changing Hill constant at low tensions that improves saturation under those conditions. They also pointed out that during most diving, which is probably aerobic (see Chap. 13), such low tensions do not develop.

9.2.2 Effect of Submersion on Blood Gases

To determine changes in blood gases during diving, Wood et al. (1984) followed a tethered, but vigorously swimming green sea turtle to a depth of 20 m to draw a blood sample. Despite the compression effects on lung gas tension the arterial blood was only 45% saturated, which indicated a large right to left shunt. In an easier approach to determining changes in blood gas tensions during breath-holds, measurements were made during resting submergences in the freshwater turtle, *P. Scripta* (Burggren and Shelton 1979). During short resting submergences of 5 to 15 min there was a steady decline in Pa_{O_2} from about 8.0 kPa down to 4.0 kPa, while PA_{O_2} fell from about 16.0 to 12.0 kPa. In contrast,

111

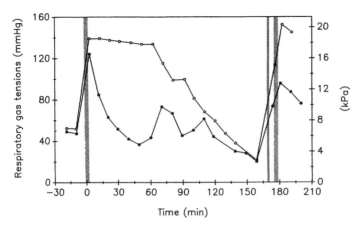

Fig. 9.2. P_{O_2} in femoral artery blood (*closed circles*) and lung gas (*open circles*) during extended dives of a 1.2-kg freshwater turtle, *Pseudemys scripta. Shaded bars* are periods of ventilation. (After Burggren and Shelton 1979)

during a long submergence of 2 h 45 min there were occasional rises in Pa_{O_2} when lung perfusion increased (Fig. 9.2). The dive ended when lung and arterial blood O_2 matched at about 2.7 kPa. In the course of these submersions the source of nearly 90% of the O_2 was from the lung.

Low Pa_{O_2} of about 5.3 kPa during ventilation are common in seasnakes as well. The P_{50} for seven species ranged from 3.2 to 6.3 kPa (T = 25 °C, pH = 7.0 and P_{CO_2} = 7.6 kPa), and consequently the blood O_2 saturations were only 30 to 70% (Seymour and Webster 1975). The low tensions are due to a large pulmonary shunt, which is advantageous in the seasnake because of the level of cutaneous gas exchange. When the sea snake dives to depth, little O_2 is lost to the water as the lung gas tension rises due to compression. However, the cutaneous exchange may continue to provide an avenue for CO_2 and N_2 loss.

9.3 Birds

There are no measurements of blood gases during diving, a few for resting conditions, and some for forced submersions. In several species of penguins resting Pa_{O_2} = 10.0 to 11.3 kPa (Kooyman et al. 1973 b; Millard et al. 1973, Mangalam and Jones 1984). Resting $P\bar{v}_{O_2}$ = 5.9 kPa. At these arterial and venous tensions the blood is 98 and 75% saturated, respectively (Fig. 9.1).

9.3.1 Hemoglobin Affinity for O_2

Compared to mammals and reptiles the avian dissociation curve is shifted to the right (Fig. 9.1). However, there has been some dispute about the high P_{50}. The argument was that the nucleated erythrocytes of birds have a high O_2 consumption and require special handling to obtain an error-free measurement (Lutz et al. 1973). However, later experiments did not verify this argument (Scheid 1979).

9.3.2 Effect of Submersion on Blood Gases

At the end of 2 min of forced submersion, the Pa_{O_2} of both the domestic mallard and a cormorant, *Phalacrocorax auritus* species, had fallen to about 5.3 kPa, and the Pa_{CO_2} had risen from 4.7 kPa to a range from about 7.3 to 10.0 kPa (Mangalam and Jones 1984). In an attempt to discriminate the effectiveness of reduced blood flow on conservation of body O_2 stores, Mangalam and Jones (1984) ventilated individuals of these two species on 100% O_2 before the dives. This raised the blood O_2 reserves to a level at which only dissolved O_2 was consumed during the dive. Those birds with lower heart rates had less of a drop in Pa_{O_2}.

In the course of 3 min of forced submersion of the gentoo penguin, Pa_{O_2} declined from about 10.7 kPa to about 4.0 kPa (Kooyman et al. 1973 b; Millard et al. 1973). At the latter tension the blood was about 25 to 30% saturated. The

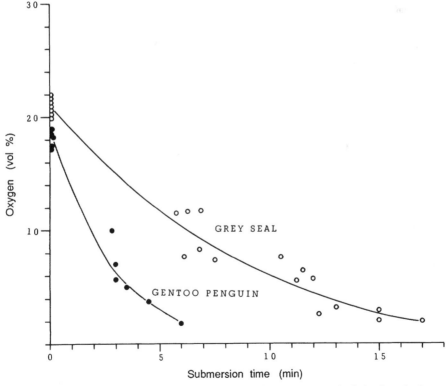

Fig. 9.3. Arterial oxygen content of a young gray seal and adult gentoo penguin during forced submersions. The points are based on results from Scholander (1940). (Kooyman 1975)

113

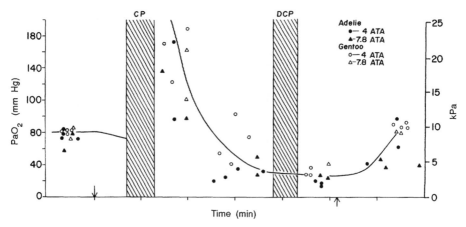

Fig. 9.4. Arterial O$_2$ tensions in Adelie and gentoo penguins during forced submersion and compression to simulated depths of 30 and 68 m. (After Kooyman et al. 1973 b)

high rate of O$_2$ desaturation of the penguin compared to the longer diving gray seal is shown in a comparison of the two during forced submersions (Fig. 9.3).

When penguins were forcibly submerged to simulated depths of as much as 68 m, the exchange continued between the air sacs and the lungs (Kooyman et al. 1973 b), and Pa$_{O_2}$ at least doubled to 24.0 kPa during the early part of compression (Fig. 9.4). Thereafter, it decreased rapidly down to about 4.0 or 5.3 kPa (Fig. 9.4). These patterns of gas tension change with breath-holding and pressure are similar to those that have been determined in seals, but with a shorter time constant.

9.4 Mammals

Resting Pa$_{O_2}$ of marine mammals (seals) is not the same as those previously mentioned for terrestrial mammals because of wide oscillations due to apneustic breathing in which the breath-hold may last for several minutes. During breath-hold Pa$_{O_2}$ may decline from about 11.3 kPa to 3.3 kPa (Fig. 9.5). Concurrently the Pa$_{CO_2}$ rises from about 6.7 to 7.3 kPa. When Pa$_{O_2}$ = 3.3 kPa the blood is about 50% saturated (Fig. 9.1).

9.4.1 Hemoglobin Affinity for O$_2$

Figure 9.1 shows that the dissociation curves vary; two of the most striking are the manatee and sea otter, both of which probably rely extensively on lung O$_2$ stores. In the Florida manatee P$_{50}$ = 2.2 kPa (pH = 7.4, P$_{CO_2}$ = 5.3 kPa, T = 37 °C) and in the sea otter P$_{50}$ = 4.1 kPa (pH = 7.4, P$_{CO_2}$ = 5.3 kPa) (Fig. 9.1).

114

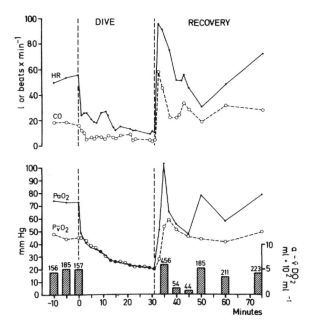

Fig. 9.5. Selected physiological variables measured during a 31-min forced submersion of a Weddell seal. *HR* heart rate; *CO* cardiac output; *number over bars* the calculated O_2 consumption in ml O_2 kg^{-1} h^{-1}, and the bars are the a–v̄ difference. (From Qvist unpubl.)

9.4.2 Effect of Submersions on Blood Gases

During resting apneusis, it seems likely that blood flow remains open to meet the perfusion needs of the various organs, as proposed in Chapter 6. The ventilatory tachycardia and increased cardiac output is mainly to meet perfusion needs of the lung for good $\dot{V}/\dot{Q}$ matching and rapid O_2 loading and CO_2 unloading of Hb and Mb. If so, then the rate of decline in Pa_{O_2} during apneusis should be greater than in forced submersions during which blood flow is very restricted. However, due to the oxygen dissociation properties of hemoglobin, the tensions decrease rapidly to the level at which there is little change in tension with a large change in content (Fig. 9.1). That range is between 5.3 to 2.7 kPa. To discriminate more effectively by means of Pa_{O_2} whether perfusion is more or less broad, the animals would need to be ventilated on 100% O_2, as done by Mangalam and Jones (1984), as discussed above.

In the course of submersion, arterial and venous O_2 content and tension quickly become equivalent. In the only experiment to my knowledge in which pulmonary arterial blood O_2 tensions (mixed venous tension) were measured at the same time as aortic blood O_2 tension, it was found in the Weddell seal that after about 2 min of forced submersion, Pa_{O_2} had dropped to 5.3 kPa and equaled $P\bar{v}_{O_2}$. Throughout the rest of the 30-min dive they remained equal as both steadily declined to about 2.7 kPa (Fig. 9.5). This indicates: (1) little exchange and added O_2 from the lung, (2) broad mixing of arterial blood and the central venous store resulting in equal O_2 tensions, and (3) venous content closely approximates arterial content during most of the submersion. This is somewhat similar to observations in the northern elephant seal in which O_2 content of ab

115

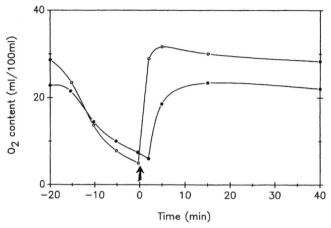

Fig. 9.6. Oxygen content in blood from abdominal aorta (*open circles*) and inferior vena cava (*closed circles*) of a northern elephant seal during a forced submersion of 20 min. Dive ends (*arrow*) at time 0. (After Elsner 1969)

dominal aorta blood and inferior vena cava blood became equal about 6 min after submersion. Remarkably, by 15 min O_2 content in inferior vena cava exceeded that of abdominal aortic blood (Fig. 9.6).

9.4.3 Diving

During natural dives as the diver usually descends to depth, hydrostatic pressure compresses the lung and raises the partial pressures of all gases. Similar to the Pa_{O_2} mentioned earlier for birds, in the seal Pa_{O_2} it increases quickly to over 26.7 kPa and then falls rapidly to 2.4 to 4.0 kPa over the remainder of the 10- to 30-min dive (Fig. 9.5). Such a low O_2 tension at the end of relatively short dives of 10 min suggests either a high metabolic rate, an open circulation, or both.

End-tidal gas samples also show this pattern of arterial O_2 tensions at the end of even short dives. During foraging dives of sub-adult Weddell seals, of a mean dive duration of only 4.9 min, end-tidal O_2 tension = 3.7 kPa (R = 2.1–5.3 kPa; N = 7, E.A. Wahrenbrock unpubl.). Similarly, end-tidal (see Chap. 4 for discussion why samples assumed end-tidal) $P_{O_2} \cong 2.4$ kPa were measured in bottle-nosed dolphins trained to breath-hold on command for 4 min, or while swimming at 20 m for 2.5 min (Ridgway et al. 1969).

In the two examples in which the dolphin and seal were diving, low PA_{O_2} suggests a high $\dot{M}_{O_2}$ which would cause a rapid decline in P_{O_2} due to either, (a) an open circulation throughout the dive, or (b) the depletion of muscle O_2 during about the first half of the dive and muscle Mb loading during the second half, as the diver ascends and an anticipatory increase in cardiac output and blood flow distribution occurs. The latter pattern agrees readily with the model proposed in Chapter 8 (Fig. 8.10).

Incidentally, although the end-tidal of 2.4 kPa observed in the dolphins is low, if it is close to Pa_{O_2}, then the blood is still 30% saturated (Fig. 9.1). At that

116

level of saturation there seems no need to invoke the high degree of anaerobicity implied by the authors (Ridgway et al. 1969), and which is no lower than some measurements in seals in which it is known that they are still within their aerobic limit, i.e., no increase in blood LA after the dive.

9.4.4 Fluctuations in Concentration of Hemoglobin

During dives, blood O_2 content appears to be enhanced (see also Chap. 5), at least in some seals. In Weddell seals there is a marked difference in hemoglobin concentration between pre-dive sample averages of 17.4 g %, and post-dive averages of 22 g % (Kooyman et al. 1980). The increase in Hb concentration occurs within the first 12 min of the dive (Qvist et al. 1986; Fig. 5.1). This maintains O_2 saturation at 20 vol % for the first 12 to 17 min of a dive in the face of a declining Pa_{O_2} (Fig. 9.7). However, the constancy of O_2 content during such single dives is probably not a general pattern of dive bouts. This was discussed earlier in Chapter 5, where it was shown that Hct rises during the beginning of a dive bout and remains elevated throughout the bout (Fig. 5.2). This is somewhat in contrast to the interpretation of Qvist et al. (1986), in which they proposed that the increase and decrease in PCV and Hb occur during each dive. From their report it seems likely that their data were obtained from animals either at the beginning of a dive bout, or ones that were making intermittent dives with long, 10 – 15 min or greater, recoveries. This is also the probable reason for the inter-

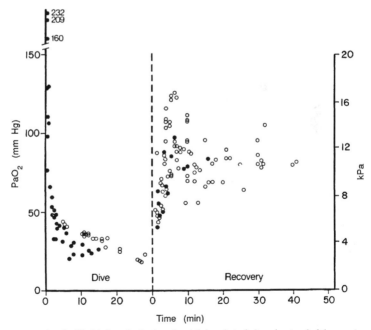

Fig. 9.7. Arterial oxygen tension in Weddell seals during short (*closed circles*) and extended (*open circles*) dives, and during the recovery. (After Qvist et al. 1986)

pretation that the spleen, the proposed source of extra red blood cells, may behave as a significant extra O_2 source for each dive, the so-called "scuba tank" effect (Zapol 1987; Hochachka and Guppy 1987). The only time that this effect might occur is during the first few dives of a series or bout, or during the rare extended dive. Rather than a scuba tank analogy, in which the tank is a separate and large O_2 store that is replenished before each dive, it might be a closer simile to "blood doping" in which a large volume of red blood cells are injected into the circulation and remain there for the entire episode of activity.

The source of Hb for the increase is unknown, but the most conventional explanation, i.e., that proposed by Qvist et al. (1986), is that the extra cells come from the spleen. In exercising terrestrial mammals the spleen may release enough erythrocytes to increase the hematocrit substantially (see Chap. 5). Indeed if aerobic diving has similarities to conventional exercise in terms of circulatory requirements of organs, then much of the increased O_2 transport resulting from the increased hematocrit may be critical if mesenteric and renal blood flow are sustained while diving, as suggested in Chapter 7. This reasoning is drawn from the observation by Vatner et al. (1974) that vigorously exercising dogs maintained mesenteric and renal blood flow as long as circulating blood capacity could be enhanced by splenic contraction. When this was denied by splenectomy the mesenteric and renal resistance to blood flow while exercising increased 144 and 199% respectively.

9.5 Conclusions and Summary

There are only limited data on tension of blood gases under resting conditions for aquatic reptiles and birds, and only a few measurements during forced submersions. Even in mammals the data are not extensive. Rarest of all are the simultaneous measurements of arterial and mixed venous blood gas tensions and contents. It has been shown that arterial and venous blood tensions become equal soon after forced submersion begins. Measurements of blood gases during diving have been obtained only from Weddell seals. The results showed patterns somewhat similar to those for forced submersions, with one important exception. The Hb content of the arterial blood increases steadily for the first 10 min of the first dive of a bout. The source of the Hb is unknown, but by convention from studies of other mammals it is most likely the spleen. Whether the change in concentration of Hb is peculiar only to Weddell seals, or is a response in other divers as well is unknown.

Chapter 10

Metabolism

The means by which divers achieve breath-hold durations of foraging and prolonged dives are not understood. Emphasis on possible regulators of the breath-hold often reflects the bias of the author. My concern has been primarily with oxygen. In Chapter 4 I reviewed the quantity of the store, and in Chapter 8 its utilization by muscle. Now I will discuss overall consumption of oxygen ($\dot{M}_{O_2}$), also expressed as the metabolic rate (MR), or frequently converted to total energy metabolism. My special interest in oxygen is its relationship to the breath-hold.

Oxygen is perhaps not the primary stimulus to breaking the breath-hold, but it is the main, short-term resource of the animal, the utilitization and depletion of which has a cascade effect on the amount and rate of increase in CO_2, and decrease in pH, glucose, and metabolism. This chapter will deal first with metabolic rate, then hypometabolism, and conclude with the types of fuel.

10.1 Metabolic Rate

Basal metabolic rate has limited intrinsic interest to the comparative physiologist for the reason expressed so well by G. A. Bartholomew (1977) that BMR is "based on physiological states that are rarely, certainly reluctantly, and perhaps never, attained by animals living under natural conditions". To meet the criteria of BMR the animal must be post-prandial, not growing, not breeding, stationary, and within its thermoneutral zone. In addition, according to Aschoff and Pohl (1970), the time within the circadian cycle and the intensity of light are also important. Because these conditions have been rarely met in the course of studies of marine reptiles, birds, and mammals, I have resorted to the broader term of resting metabolic rate (RMR). In my use this is the MR when the animal is post-prandial, stationary, and for endotherms presumably, but not necessarily proven, in its thermoneutral zone.

The RMR is a useful standard to help assess the possibility of hypometabolism during dives, and to determine the metabolic scope of swimming and of hunting dives. It has been measured in a variety of marine birds and mammals of a wide range in size. Empirically some of the values determined seem to come close to the rate predicted from scaling equations. However, it is not my objective to enter the controversy on scaling and appropriate exponents, but rather to use it as an empirical tool for standard comparisons, the calculated values of which are derived from equations familiar to every vertebrate biologist.

10.1.1 Reptiles

10.1.1.1 Resting Metabolic Rate

The RMR of reptiles was reviewed extensively by Bennett and Dawson (1976); however, this review contains little information on marine forms. In the same year the RMR of green sea turtles over a size range of 0.03 to 142 kg was reported (Prange and Jackson 1976). Over this range the RMR in O_2 l min^{-1} was:

$$\dot{M}_{O_2} = 0.00097 \ kg^{0.826} \tag{10.1}$$

The temperature of this estimate was unclear, but appeared to be about 25 °C. As noted below, this RMR is an order of magnitude less than birds and mammals at body temperatures of 37 to 40 °C.

10.1.1.2 Metabolic Rate While Swimming

The only reported metabolic rate (MR) determined during swimming is that of 0.74 kg green sea turtles (Prange 1976). At rest $\dot{M}_{O_2} = 1.2$ ml kg^{-1} min^{-1}, and at maximum speed of 0.35 m s^{-1} $\cong$ 4.2 ml kg^{-1} min^{-1}. The RMR was 17% above that predicted by Eq. (1), and the maximum $\dot{M}_{O_2}$ was four times RMR. In comparison, the maximum $\dot{M}_{O_2}$ of a 142 kg green sea turtle was ten times RMR while it crawled for 20 min over a grassy field (Prange and Jackson 1976). The latter value is remarkably high for reptiles.

10.1.2 Birds

10.1.2.1 Resting Metabolic Rate

RMR of a large number of avian species has been measured; scaling relationships have been studied extensively. Two of the most detailed analyses have been done by Lasiewski and Dawson (1967) and Aschoff and Pohl (1970). In the latter summary for nonpasserine birds the following equations were, respectively, derived:

$$\dot{M}_{O_2} = 0.0113 \ kg^{0.723} \tag{10.2}$$

$$\dot{M}_{O_2} = 0.013 \ kg^{0.729}, \text{ for active phase} \tag{10.3}$$

$$\dot{M}_{O_2} = 0.010 \ kg^{0.734} \ \varrho, \text{ for quiet phase} \tag{10.4}$$

where $\dot{M}_{O_2}$ is in l of O_2 per min. In the course of reviewing these and other reports it has been necessary to convert from other units to l of O_2. I have used the factors of Bartholomew (1977) which assumes an RQ of 0.79. This may cause a slight error in those animals whose metabolic fuel is mainly fat. For example, the RQ determined in Humboldt penguins was 0.70 (Drent and Stonehouse 1971) and 0.73 in emperor penguins (Pinshow et al. 1976). Equation (3) will be subsequently used to calculate RMR to compare with values obtained in the following review of reports on swim metabolism in ducks and penguins.

120

Table 10.1. Resting and swimming metabolic rates in marine birds, using Eq. (10.3) $M_{O_2} = 0.013 \text{ kg}^{0.729}$ where M_{O_2} is l of O_2 min^{-1} and divided by body mass to obtain the predicted M_{O_2} in proportion to body size

Species	Mass	Resting M_{O_2}	Predicted M_{O_2}	H_2O or air temperature	Swim M_{O_2}	Speed	Reference
	(kg)	(ml O_2 kg^{-1} min^{-1})		°C	(ml kg^{-1} min^{-1})	(m s^{-1})	
Eudyptula minor	1.2	19	12.4	19–22	25	0.72	1
Spheniscus humboldti	3.78	10.8	9.1		37	1.25	2
Spheniscus humboldti	4.6	13	8.6	18	16	?	3
Spheniscus humboldti	3.87	7.3	9.0	10–25[a]		–	3
S. demersus	3.17	8	9.5		80	?	4
Aptenodytes forsteri	23.4	5.5	5.5	−10 to 20[a]	–	–	5
Aythya fuligula	0.597	16.3	8.9		56.6	–	6

[a] Air Temperature.
1) Baudinette and Gill (1985).
2) Hui (1983).
3) Butler and Woakes (1984); Drent and Stonehouse (1971).
4) Nagy et al. (1984).
5) Pinshow et al. (1976).
6) Woakes and Butler (1983).

10.1.2.2 Metabolic Rate While Swimming

By means of high speed O_2 analysis with a mass spectrometer, and multiple regression analysis of the results of O_2 concentration, dive duration and surface duration the MR of dives in tufted ducks was determined (Woakes and Butler 1983). $\dot{M}_{O_2}$ was near maximum during dives averaging 14.45 duration. This was equivalent to 3.5 times RMR (Table 10.1), which was nearly two times that predicted by Eq. (1). It is perhaps significant that the water temperature was 13.6 °C, and that the ducks may not have been thermally neutral.

Penguin RMR, and swimming MR at the maximum velocity measured are summarized in Table 10.1. In two instances, Humboldt and jackass penguins, the swim velocities are unknown, but the circumstances indicate that most of the time swimming was probably at the preferred velocity. Such a swim rate is most likely below the maximum. Some of the RMR's are high according to Eq. (3). The little blue penguin is about 50% above predicted values, as is the single measurement for a Humboldt penguin. Swimming metabolism was two to four times predicted RMR, but was up to nearly nine times in one study of Humboldt penguins in which diving MR was estimated from time-partitioning analysis.

10.1.3 Mammals

10.1.3.1 Resting Metabolic Rate

The RMR of mammals has been analyzed and scaled more extensively than for all other vertebrate groups. Any textbook in comparative physiology presents the so-called mouse to elephant metabolic curve, and in fact it has been extended to such a degree that it might be more appropriately called the shrew to whale curve, although the estimates of MR of large whales are shaky. Such a curve covers about seven orders of magnitude of mass. The equation from Kleiber (1961) is:

$$\dot{M}_{O_2} = 0.0101 \text{ kg}^{0.75}. \tag{10.5}$$

Perhaps a more familar variation is from Schmidt-Nielsen (1983):

$$\dot{M}_{O_2} = 0.0113 \text{ kg}^{0.75}. \tag{10.6}$$

In general the measured RMR of marine mammals is usually about 1.5 to 3 times that predicted from the above equations. The difference may seem even greater if compared to a lean man whose body fat is about 10% of total body mass (Astrand and Rodahl 1977), compared to about 40% for marine mammals. Nevertheless, another perhaps comparably low metabolizing tissue is bone, for which in the male southern elephant seal, *M. leonina*, it is 6–8% of total mass (Bryden 1972) compared to 15% for the human male (Kleiber 1961). Overall the two masses of blubber and bone may balance out, eliminating a real difference in RMR.

That there is no difference in RMR is the conclusion from a recent survey by Lavigne et al. (1982, 1986), who reviewed extensively the results of a variety of marine mammals. They noted that most of the marine mammal studies of RMR do not fit the criteria of RMR stated above. In those reports in which conditions closely conformed to the criteria of RMR and time was taken to permit the animals to adjust to the metabolic chamber, the RMR was close to that predicted by the equations. Hence, I use Eq. (5) for comparison of RMR and swimming metabolic rates obtained in the studies discussed below.

10.1.3.2 Metabolic Rate While Swimming

Most studies of swimming metabolism have not controlled speed. These in brief were (1) a sea lion swimming around a tank and on cue breathing into a collecting bag (Costello and Whittow 1975), (2) a harbor seal swimming below a covered pool and breathing into a collecting valve (Craig and Pasche 1980), (3) Weddell seals hunting below antarctic fast ice, but required to return to a single breathing hole where expired gas samples were collected (Kooyman et al. 1973 a), and (4) harbor seals that were required to swim upright in a deep cylinder in which swim effort was controlled by adding or removing known weights from a vest worn by the seals (Elsner and Ashwell-Erickson 1982). In passing, some MR's have been estimated recently in large whales by the rate of ventilation (Dolphin 1987; Sumich 1983).

In all cases stated above RMR was greater than the predicted resting $\dot{M}_{O_2}$. The greatest increase over the measured RMR was 3.7 times in the weighted harbor seals. If that $\dot{M}_{O_2}$ is compared to the predicted RMR the peak $\dot{M}_{O_2}$ is then 7.3 times greater. The authors considered this near the $\dot{V}_{O_2}$ max of harbor seals. However, my colleagues and I have obtained values somewhat higher.

To control swimming speed and determine the relationship between velocity and $\dot{M}_{O_2}$, harbor seals were allowed to swim in a water channel in which water flowrate could be varied from 0.5 to 1.4 m s^{-1} (Davis et al. 1985). The harbor seals' $\dot{M}_{O_2}$ increased curvilinearly with velocity. In young seals of average mass of 33 kg the equation was:

$$\dot{M}_{O_2} = 5.1 + 6.25\,(\dot{V})^{1.42} \tag{10.7}$$

where $\dot{M}_{O_2}$ = ml O_2 kg^{-1} min^{-1} and $\dot{V}$ = m s^{-1}. In an adult 63 kg seal the equation was:

$$\dot{M}_{O_2} = 4.6 + 3.1\,(\dot{V})^{1.42} \tag{10.8}$$

Resting rates in this apparatus were 20 to 30% above predicted levels (Table 10.2) and at 1.4 m s^{-1} $\dot{M}_{O_2}$ had increased to 2.5 and 3.6 times the predicted RMR.

Similar measurements have been obtained for young California sea lions in which maximum velocity was 2.3 m s^{-1} (Feldkamp 1985). The best fit equation for the relationship was:

$$\dot{M}_{O_2} = 6.77e^{.48\ \text{vel}} \tag{10.9}$$

The measured RMR in the sea lion was about 40% above the predicted value. Much care was taken to bring the sea lion into a resting state, and considering the high level of activity characteristic of sea lions, this result may be as low as one can expect to achieve, and much lower than the value of Costello and Whittow (1975) in which the measured RMR was four times the predicted RMR. This latter value is an even greater multiple of RMR than that of Weddell seals floating in $-2\,°$C sea water (Table 10.2).

The resting and diving MR of Weddell seals are the only values that have been obtained for a free-ranging and hunting animal. Originally I thought that some of the elevated recovery MR (O_2 debt) was not repaid until later in the dive

Table 10.2. Resting and swimming metabolic rates in marine mammals. Equation (10.6) was used to calculate predicted M_{O_2}. $M_{O_2} = 0.0101\ \text{kg}^{0.75}$, where $M_{O_2} = 1$ of O_2 min^{-1}

Species	Mass	Resting M_{O_2}	Predicted M_{O_2}	H$_2$O Temp.	Swim M_{O_2}	Speed	Reference
	(kg)	(ml O_2 kg^{-1} min^{-1})		°C	(ml kg^{-1} min^{-1})	(m s^{-1})	
Phoca vitulina	33	5.1	4.2	15–18	15	1.4	1
Phoca vitulina	63	4.6	3.6	15–18	9	1.4	1
Phoca vitulina	25	8.8	4.5	15–18	32.8	?	2
Leptonychotes weddellii	425	5.2	2.2	− 2	4.2	?	3
Zalophus californianus	62	14.5	3.6	?	23	1.7	4
Zalophus californianus	63	6.5	4.6	26	25	2.8	5
Zalophus californianus	18			18	33	2.8	5

1) Davis et al. (1985).
2) Elsner and Ashwell-Ericson (1982).
3) Kooyman et al. (1973 a).
4) Costello and Whittow (1975).
5) Feldkamp (1985).

bout, or at the end of the bout. However, in view of the results on [LA] discussed in Chapter 8, it now appears that the only O_2 debt of these aerobic dives is the loading of depleted O_2 stores (Kooyman et al. 1980). This is accomplished during the surface interval between dives so that the post-dive MR is an accurate estimate of the dive MR and not an erroneously low value. It is the high RMR obtained after the dive bouts that are not understood.

Using the post-dive RMR as a comparison, these results have been mis-construed at times to be evidence for hypometabolism during the dive because the dive MR is lower than the post-dive RMR (Kooyman et al. 1980; Guppy et al. 1986). It seems likely that the elevated RMR was due to increased visceral metabolism while digesting the meal, and in particular the result of the specific dynamic action (SDA) of protein metabolism. For example, it has been shown in sea otters that within 30 min after a meal MR rises to 54% above resting MR, and remains elevated for about 3 h (Costa and Kooyman 1984). This is about the time period in which we measured the MR in resting seals. The following discus-sion exposes further uncertainties in the proposition that hypometabolism occurs during diving.

10.2 Hypometabolism

This is a rate of metabolism lower than that which occurs under the standard conditions of resting in the post-prandial and normally quiet period of the 24 h cycle. In endotherms it is also within the animal's zone of thermoneutrality.

10.2.1 Reptiles

The tolerance of reptiles, particularly freshwater turtles, to anoxia was reviewed in Chapter 3. It is clear that much of their breath-hold capacity is due to a reduced body temperature and hypometabolism (Herbert and Jackson 1985). Such a low level of metabolism cannot be matched in birds and mammals. Even at rest the MR of birds and mammals is an order of magnitude higher than that of reptiles [see Eqs. (1), (2) and (6)]. Nevertheless, the acceptance that some degree of hypometabolism occurs and benefits the breath-hold capacity of aquatic birds and mammals is indicated by the uniform concurrence expressed in all textbooks that discuss adaptations to diving. However, the evidence therefore is limited and indirect, and understandably so because the study of reduced metabolism during the nonsteady state condition of diving is intractable.

10.2.2 Birds

There have been two studies on birds, both on the domestic mallard; this enig-matic nondiving dabbler out-performs and has been more thoroughly studied than any other "diver". Andersen (1959) suggested a reduced metabolism from evidence of a decreased rate of depletion of O_2 from the trachea, and a small drop in rectal temperature during the dive. Rectal and abdominal temperature fell even further for about 60 min after the dive. This continued drop was not explained, and considering the nonsteady state condition of the dive, neither the temperature nor gas measurements make a strong case for reduced metabolism.

Direct calorimetry during the dive of a mallard was later attempted by Pickwell (1968). Since the approximately 2.5-kg ducks had such a large heat capacity, only a small heat loss occurred during the dives. It was about 2% of total heat content and appears as only about a 5 to 15% decline in metabolism. To amplify the signal, heat loss was compared to a recently dead duck, at zero metabolism. Based on this comparison the estimates of reduced metabolism of the diving duck were as much as 95% below the pre-dive level. Since no similar subtractions were made from resting ducks used as controls, it seems possible that the estimates of reduction in heat production may be high.

Some of the earliest work on the problem of hypometabolism was based on the comparison of O_2 deficit during the dive and O_2 debt after the dive (Scholander 1940). The oxygen deficit is the expected amount of O_2 consumed during the dive minus the actual amount consumed. The shortfall from aerobic energy needs is made up of anaerobic energy-yielding processes. The O_2 debt is the total aerobic metabolism less the resting metabolism after the exercise or dive period. Under severe exercise conditions the O_2 debt is usually double the deficit because of the additional O_2 consumption necessary for the resynthesis of LA to glycogen (Astrand and Rodahl 1977). Yet often the O_2 debt measured after a dive resulted in a value much lower than the deficit (Scholander 1940).

10.2.3 Mammals

To make valid measurements it must be assumed that the O_2 repayment must be immediate and not protracted over an extended period, which would result in a subtle difference in resting $\dot{M}_{O_2}$ and elevated recovery $\dot{M}_{O_2}$, and the resting $\dot{M}_{O_2}$ selected for the comparison is constant both before and after the dive. Assuming that these criteria prevailed during seven nonstruggling forced submersions, the O_2 debt was from 17 to 50% of the deficit. In struggling forced submersions the debt ranged from 30% to greater than the deficit. However, there is a degree of uncertainty about these experiments because of the level of pre-dive $\dot{M}_{O_2}$. In *Cystophora cristata* (29 kg) it was 7.6 ml kg^{-1} min^{-1} (N = 2, 1 animal) and in *Halichoerus grypus* (39 kg, N = 3, 3 animals) it was 9.0 ml kg^{-1} min^{-1}. These $\dot{M}_{O_2}$ rates are 2 to 2.5 times the predicted RMR of Eq. (6). Nevertheless, the results reflect a consistent pattern of an unbalanced O_2 debt, which prompted another experiment to test the hypothesis of reduced metabolism

The temperature of the brain and other areas of the body were monitored in young Harbor seals during 15-min dives (Scholander et al. 1942). The brain temperature dropped as much as 2.5 °C and in other areas such as muscle, liver, and abdomen, the temperatures decreased also, but not as much as the brain. In most instances similar to the duck mentioned earlier, temperatures continued to decrease for about 30 min after the dive. Also similar to the duck there was little explanation for this phenomenon. In such a nonsteady state condition as a forced submersion it seems that the temperature changes are only qualitative indicators of heat production, particularly in view of the fact that there is a continued fall after the dive in all areas measured, even though MR must have increased. Later forced submersion experiments on young harbor seals have perhaps confounded this issue somewhat. Although rectal temperatures fell 1 or 2 °C there was little or no change in hypothalamic temperature (Elsner et al. 1975).

Declines in body temperature are not limited to nonexercising seals during forced submersions. The decrease in core temperature also occurs during voluntary diving. During a sequence of short duration dives of adult Weddell seals the blood temperature decreased rapidly to 36.6 °C and remained at this temperature throughout the sequence (Qvist et al. 1986). Furthermore, exploratory dive body temperature was as low as 34.9 °C after a 53-min dive, and steadily increased in the course of recovery back to 38 °C (Kooyman et al. 1980). Under diving conditions, although the metabolism of some organs may be reduced, it seems unlikely that overall metabolism is, as indicated by the previous discussion of dive M_{O_2} of Weddell seals. However, another means of assessing dive MR has been attempted recently.

Based on the post-dive level of blood [LA] of 8 mM l^{-1} at the end of a 30-min dive, as well as some estimates of muscle O_2 uptake, overall MR was calculated. The result was too small to account for the proposed debt and increase in [LA] which should have been at least 21 mM kg^{-1} body mass. It was suggested that a reversed Pasteur effect may obtain (Guppy et al. 1986). The Pasteur effect, which is the increased consumption of glucose in response to depletion of O_2, is due to less complete utilization of glucose under anaerobic conditions. The weakness in the above conclusion is that blood [LA] does not reflect muscle concentration or turnover well enough for reliable calculations.

From the above discussion it is clear that there are several caveats in accepting hypometabolism as a common occurrence in diving animals. From the studies of overall metabolism in hunting Weddell seals it is apparent that overall M_{O_2} is elevated (Kooyman et al. 1973 a). However, this does not exclude the possibility that some organs may reduce metabolism during dives, especially prolonged ones. Indeed, the rising CO_2 levels in various organs during the dive may be the "smoke that dampens the metabolic fire" to quote F. N. White (pers. commun.). For example, Severinghaus (1974) reported that CO_2 has five times the narcotic potency of N_2O, and it may be the H^+ rather than P_{CO_2} that is the effector. Further, a decrease in pH causes a small decrease in $\dot{M}_{O_2}$ in dogs (Albers 1974). If so, then the rising H^+ concentration due to production of CO_2 and LA may have a potent effect on tissue metabolism in diving animals in which the effect becomes greater the longer the dive.

10.3 Fuels

Before concluding the discussion of metabolic rates and addressing its possible consequences to the aerobic dive limit, a few remarks about recent work on the fuel that sustains aquatic birds and mammals seems appropriate. In Chapter 4 reference was made to carbohydrate metabolism during forced submersions and prolonged dives and the subject of fuels was broached in Chapter 8. However, most aquatic animals do not make many prolonged dives, and furthermore consume a diet high in fat. Davis (1983) pointed out that the aerobic metabolism of carnivores whose diet is mainly low in carbohydrate tends to utilize fat as the primary fuel, with RQ's about 0.70. Consistently the RQ of resting marine birds and mammals is also near 0.70. However, for a diving animal, especially during a breath-hold, to rely on fat may seem a contradiction in one sense. It requires about 11% more O_2 per carbon unit for complete oxidation of fatty acid than it does for glucose (Vik-Mo and Mjos 1981), or about 6% more energy is produced per O_2 consumed for carbohydrate than fat (Schmidt-Nielsen 1983). However, this difference is only a fraction of the 40 to 25% of the "O_2 wasting effect" that occurs when the fuel for O_2 consumption of the rat or dog myocardium is changed from carbohydrate to fat (Vik-Mo and Mjos 1981). Whether there is a difference in O_2 consumption in any organ of a diving animal is unknown. However, during resting and aerobic exercise metabolism in seals the major fuel is clearly not carbohydrates.

In resting seals the combined contribution of LA and glucose oxidation contribute less than 9% to resting metabolism (Davis 1983). Much more of the LA turnover (21%) is oxidized than is the glucose turnover (3%). Furthermore, less than 27% of the LA produced during forced submersions is oxidized upon recovery. According to the reference made earlier about resynthesis of [LA] to glycogen (Astrand and Rodahl 1977), this pathway should increase the O_2 debt above the deficit rather than result in a decrease below the deficit as discussed above. Also, although there is a consistent post-submersion hyperglycemia after prolonged breath-holds (Davis 1983; Kooyman et al. 1980; Murphy et al. 1980; Robin et al. 1981) glucose oxidation is only 11% higher than pre-dive levels (Davis 1983). The high glucose concentration is due to a high glucose entry rate that reaches maximum level about 30 s after the dive, and which returns to pre-dive levels within 15 min. However, due to the low removal rate, blood glucose concentration does not return to pre-dive levels for more than 2 h.

In the steady-state condition of exercise achieved by exercising seals at known levels of swim velocity, the turnover rate of various fuels was determined (Davis et al. 1985). At a $\dot{M}_{O_2}$ of three times resting the RQ was between 0.71 and 0.74. About 27% of energy metabolism could be accounted for by oxidation of LA, glucose and FFA. At this level of metabolism LA and glucose accounted for 4% of total energy metabolism, and the FFA oxidation rate had increased over four fold from the resting level and accounted for 23% of the total energy metabolism.

10.4 Conclusions and Summary

Clearly the subject of metabolism is in foment. The conventional view of elevated metabolic rates in marine mammals has been challenged. Some of the difficulties have arisen because of nonstandard procedures, but others may be philosophical. How important is BMR or RMR in the daily life of aquatic birds and mammals, especially during the pelagic phase of their life cycles when deep sleep may be a dangerous luxury, and where, at least in the smaller forms, the environmental waters may be below their thermoneutral zone? A clearer picture may be revealed through the recent application of doubly-labeled water for metabolism and discrete time partitioning for activity.

As we argue about whether aquatic birds and mammals fit a hypothetical scaling curve, the subject of hypometabolism remains on the sidelines as an accepted axiom, but with relatively little proof. What little there is harks from unnatural experiments and tells us nothing about the levels of organ and whole animal metabolism of diving. At present, this subject has been overshadowed by studies of intermediary metabolism.

A relatively new area of research, the study of the fuels that fire divers, has made good progress in marine mammals, but few or no such studies have been done on birds and reptiles. In seals it appears that in most circumstances carbohydrate resources are conserved and lipid stores are used for most energy needs.

Chapter 11

Hydrodynamics, Swim Velocity and Power Requirements

The discussions in this chapter, which are mainly theoretical arguments of power requirements based on different schemes of drag reduction, are a continuation of some of the empirical results presented in Chapter 10. This chapter will be somewhat different from previous ones in its restriction mainly to general aspects of diving vertebrates. This is because hydrodynamic theory applies to all forms, and because there has been so little experimental work done on reptiles, birds, and mammals. Most of what is discussed below was developed from the much more extensive fish literature, as well as general hydrodynamic theory.

My perspective is that of a biologist unfamiliar with this field, who attempts to relate it to what is known about physiology and behavior of diving vertebrates, and how general concepts provide a better insight into the behavior of these animals. Also, a review should help to determine the aspects of these physical relationships that are possible to measure, how to design better exercise experiments, and to determine empirically how well some of the theoretical predictions conform with actual measurements. To me, one of the most important questions is how fast do divers normally swim, and why?

For this review, I have relied mainly on the biological literature rather than on that of the physical sciences. This is because the biologists have dealt specifically with the problems of animals rather than physical objects. My main sources have been Blake (1983 b) and Webb (1975), who dealt mostly with fish, and the more general reviews of Aleyev (1977) and Vogel (1981). As one can imagine, the more extensive studies on fish have involved small animals, so considerable extrapolations to larger animals are necessary. One distinct advantage of mammals that will be noted below is that they can be trained so that some drag values obtained only from frozen fish have been measured in trained seals and sea lions. Finally, despite more than 50 years of research, nobody has managed to measure drag of a swimming fish. This fact emphasizes the intractability of some of these basic issues.

11.1 Hydrodynamics

Over 70% of the earth's surface is covered with water to an average depth of about 4000 m. Divers are skimmers of just the upper layers. One of the most important physical characteristics of water to which aquatic animals had to adapt was its density, which is about 1000 times greater than air (Table 11.1). Also the dynamic viscosity of sea water, the internal resistance to change in form, or

Table 11.1. Some physical constants of air and sea water

	Air	Sea water	Temperature
Density (kg m^{-3})	1.29	1.028×10^3	0
		1.027×10^3	10
	1.21	1.024×10^3	20
Dynamic viscosity (kg m^{-1} s^{-1})	1.7×10^{-5}	1.92×10^3	0
		1.39×10^3	10
	1.8×10^{-5}	1.07×10^3	20
Kinematic viscosity (m^2 s^{-1})	1.3×10^{-5}	1.87×10^{-6}	0
		1.35×10^{-6}	10
	1.5×10^{-5}	1.044×10^{-6}	20

"stickiness" (Vogel 1981) is about 110 times greater than air, at 0 °C. Due to a substantial influence of temperature on water that does not occur in air, viscosity is reduced to about half (60 times) at 20 °C (Table 11.1). However, it is a relatively small factor compared to others. The joint influence of dynamic viscosity and density is expressed as a ratio (u/p) called kinematic viscosity, where u is viscosity and p is density. An intuitive surprise is that kinematic viscosity is only 14 times greater in *air* than water at 20 °C, and 7 times more at 0 °C. This viscosity is a measure of how easily a fluid flows uniformly without tending to develop vortices. These relationships have important implications in regard to the resistance of a body to flow, which is generally expressed as drag, or the antinym thrust. The following emphasis will be on the former.

11.1.1 Drag

Total drag of a body of revolution (D_T), is the sum of profile and induced drag. Here I deal only with profile drag, which is often separated into pressure or form drag (D_p), and friction drag (D_f). The former is dependent upon the shape of the object, and location of boundary layer separation. Friction drag is dependent more on the extent and characteristics of the boundary layer. This layer is discussed in detail by Vogel (1981) and I do not do the boundary layer justice by briefly defining it as the local velocity gradient that is perpendicular from the surface of the body to an outer limit where the difference from the general environment is <1%.

The importance of shape is illustrated in Fig. 11.1, which shows that a sphere induces a large pressure drag because of an early separation of flow resulting in a large wake, and a lower pressure behind relative to the head of the body. The pressure drag force is large compared to the shear stresses of frictional drag. In contrast, a streamlined body has a low pressure drag since much of the pressure on the leading face of the body is regained on the taper before flow separation occurs. However, because of the increased surface area, the frictional drag is greater than that for a sphere of similar diameter. The sum of the pressure and frictional

Fig. 11.1. Flow pattern and wake size around a sphere and streamlined body. (After Vennard and Street 1976)

drag of a streamlined body is many times less than that of a sphere at a similar Reynolds number (Re).

Theoretically, bodies of similar shape will have similar flow properties at equivalent Re. It is expressed as:

$$Re = LV/v, \tag{11.1}$$

where $v = \mu/\varrho$, the kinematic viscosity referred to above, and L, in m, is the length of the body moving at velocity (V), in m^{-1}.

Drag can be expressed by the equation;

$$D = 1/2 \, \varrho \, AV^2C_D, \tag{11.2}$$

where D is in kg, ϱ is density (kg m^{-3}), A may be either frontal area (A), wetted surface area (A$_W$), or volume$^{0.66}$ (A$_V$), and C$_D$, which is the nondimensional frontal drag coefficient (C$_{D_F}$), wetted surface area (C$_{D_A}$), or volume$^{0.66}$ (C$_{D_V}$). C$_D$ is not a constant and is related to the inertial and viscous ratio (Re) of the fluid and moving body. For a flat plate where flow is parallel to the long axis, most drag is due to friction. The C$_D$ for this body is equal to:

$$C_{D_{lam}} = 1.3 \, Re^{-0.5} \tag{11.3}$$

for a laminar boundary layer, and as:

$$C_{D_{turb}} = 0.072 \, Re^{-0.2} \tag{11.4}$$

for a turbulent boundary layer (Webb 1975). These relationships are shown graphically in Fig. 11.2 as well as that of optimum profiles. As you will see, many marine birds and mammals are close to this optimum profile and should fall near that line. Note that C$_D$ becomes constant at high Re. Profile drag is most important at high Re; the flat plate turned on edge has a C$_D$ 275 times greater than the same plate parallel to flow at Re $= 10^5$, whereas it is only 1.5 times greater at Re $= 10$.

With good profiles the C$_D$ of bodies of revolution must incorporate a shape factor into the equation. This includes both skin friction and pressure coefficients (Hoerner 1965):

$$C_{D_A} = (C_f + C_p), \text{ or} \tag{11.5}$$

$$C_{DA} = C_f[1 + 1.5(d/l)^{1.5} + 7(d/l)^3], \tag{11.6}$$

131

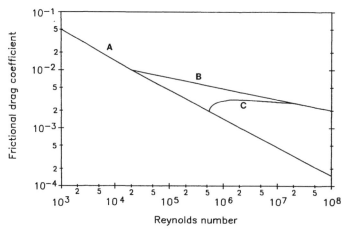

Fig. 11.2. Drag coefficients for laminar (*A*) and turbulent (*B*) boundary layer flow as a function of Reynolds number for a flat plate. *C* represents the change in drag coefficients at transitional Reynolds numbers. (After Webb 1975)

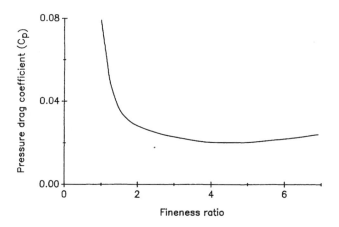

Fig. 11.3. Relationship of drag coefficient to fineness ratio in a streamlined body. (After Blake 1983 b)

where d is mean width and depth of the animal, l is body length, and C_f and C_p are frictional and pressure coefficients.

The best profiles of streamlined bodies have a fineness ratio (FR = l/d) of about 2 to 6. Ratios <2 cause the C_D to soar, and bluff (nonstreamlined, i.e., sphere) bodies tend to become turbulent at Re>10^3 (Fig. 11.3; Blake 1983 b). Based on tests of rigid models, the boundary layer beyond maximum girth becomes turbulent in all nektonic animals above Re>10^5 (Aleyev 1977).

Turbulence is not necessarily bad. Although it increases D_f, it may delay separation, and thus reduce wake size, which reduces D_p. For example, in a bluff body such as a circular cylinder with its long axis perpendicular to flow D_p = 57% of D_T at Re = 10, but 97% at Re = 10^5 (Vogel 1981). Any reduction in D_p,

132

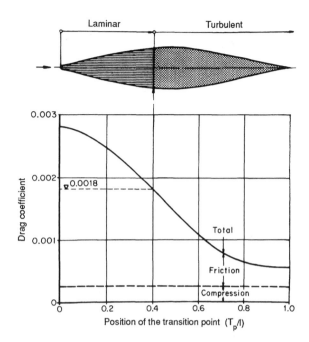

Fig. 11.4 Drag coefficient of wetted surface area of a dolphin model in relation to the position of the transition point (T_p) from laminar to turbulent flow; body length (l) = 2.2 m, velocity (V) = 10 ms^{-1} and Re = 20.6 × 10^6. (Hertel 1969)

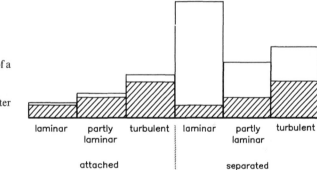

Fig. 11.5. Relative effect of various boundary layer flow conditions on the total drag of a streamlined body. *Hatched areas* are frictional drag and *clear areas* pressure drag. (After Webb 1975)

therefore, has a significant effect at high Re. This is true as well for a streamlined body, although the D_f is greater than D_p and the total of both is relatively small compared to a bluff body. Nevertheless, it is advantageous to move maximum girth back on the body. This has its limitations because too abrupt a taper may cause separation and a large wake.

The influence on C_{D_A} of the site of transition from a laminar to a turbulent boundary layer was shown graphically by Hertel (1969). The diagram shows the different C_{D_A}'s for a dolphin model dependent on where the transition point occurs (Fig. 11.4). This model illustrates only the altered effects on C_{D_A} due to C_f, while C_p remains constant. Webb (1975) carries this further with his non-dimensional comparison of turbulent flow that remains attached compared to laminar flow that separates (Fig. 11.5). Note that even turbulent but separated

133

Table 11.2. Drag coefficients determined from glides, or when towed

Species	Mass	Velocity	Drag coefficient		Re	Reference
	(kg)	(m s^{-1})	C_{D_F}	C_{D_A}		
Sphere	–	–	0.47	0.118[a]	10^5	Vogel (1981)
Flat plate (laminar)	–	–	–	0.0013	10^6	Vogel (1981)
(turbulent)	–	–	–	0.0045	10^6	Vogel (1981)
Streamlined body (lam)	–	–	0.020[b]	0.0016	10^6	Hoerner (1965)
(FR=4.5) (turb)	–	–	0.071[b]	0.0056	10^6	Hoerner (1965)
Glides						
Salmo	0.08	1.3	–	0.015	2.8×10^5	Webb (1975)
Pygoscelis papua	5.0	–	0.07	0.0044	10^6	Nachtigall and Bilo (1980)
Aptenodytes forsteri	30	2.2	0.038[b]	0.003	1.8×10^6	Clark and Bemis (1979)
Phoca vitulina	33	1.8	0.038	0.004	1.6×10^6	Williams and Kooyman (1985)
Zalophus californianus	38	2.4	0.046	0.004	2.9×10^6	Feldkamp (1987)
Lagenorhynchus obliquidens	91	4.2	0.061[b]	0.0048[c]	9×10^6	Lang (1975)
Sotalia guianensis	85	2.5	0.038[b]	0.003	10^6	Videler and Kamermans (1985)
Tursiops truncatus	232	1.9	0.102[b]	0.008	10^6	Videler and Kamermans (1985)
Tows						
P. vitulina	85	1.8	0.111	0.009	2.5×10^6	Williams and Kooyman (1985)
Homo sapiens	75	1.8	–	0.03	2.6×10^6	Williams and Kooyman (1985)

[a] To convert C_{D_F} to C_{D_A} of a sphere multiply by 0.25.
[b] To convert C_{D_A} to C_{D_F} of a prolate spheroid of 4 : 1 multiply by 12.8.
[c] To convert C_D for volume to C_{D_A} of a prolate spheroid of 4 : 1 multiply by 0.16. (Vogel 1981).

flow creates less drag than separated laminar flow. These differences are due largely to the increase in D_p because of separation. Thus, the advantage of turbulence of the boundary layer beyond the shoulder is greater stability and delay of separation to further downstream resulting in a smaller wake.

To place the theoretical drag coefficients in perspective, the calculated C_{D_F} and C_{D_A} of a sphere, flat plate, and streamlined body are given in Table 11.2. Also presented are the C_D's determined for animals while they are in a glide; the lowest drag condition. The sphere, which has the highest C_{D_A}, is about 100 times that for laminar flow over a flat plate. The plate has the lowest pressure drag of any form.

When the C_D of most marine birds and mammals is compared to a streamlined body of FR=4.5, it can be seen that most have laminar flow over the first 20 to 40% of the body, e.g., the harbor seal $C_{D_A}=0.004$, which falls between the $C_{D_{A_{lam}}}=0.0016$, and $C_{D_{A_{turb}}}=0.0056$ of a streamlined body. This agrees approximately with Fig. 11.4. In the towed seal, the C_{D_A} is about twice the fully turbulent C_{D_A} (Williams and Kooyman 1985). This is probably due to the wake created by the tow support strut, and to turning trim of the seal which was pulled around a circular tank. For comparison, a similar-sized person towed in the same manner had a C_{D_A} about four times that of the seal.

11.1.2 Physical Means of Drag Reduction

I have discussed the general features of body profile and streamlining, but there are several variations on this theme. Many of the schemes seem logical, but in most cases there is little experimental evidence to support theory. What follows is a brief discussion of the shape and surface properties of some divers.

11.1.2.1 Shape

Webb (1984) reviewed the shape of fishes in relationship to hunting strategies. The range varies from the lunge and grab strategy of tubular-shaped pike, to laterally compressed, slow-moving reef fish. Perhaps the more familiar forms are the trout (carangiform), and tuna (thunniform). These are high-speed, pelagic fish. For the fastest species, the body is rigid, and the peduncle constricted, or "narrownecked". Most of the propulsion is from the tail with little motion in the rest of the body. The reduction in body motion reduces turbulence and body drag. Narrow necks also reduce the amount of surface area of portions of the body that must move to create thrust from the tail, wings, or flippers, but provide little thrust of their own. Many of these propulsive surfaces are shaped to function as foils, rather than paddles, and thrust is provided in both phases of the stroke.

11.1.2.2 Surfaces

The character of the surface may also be modified. Hair covering a seal model reduces C_D at velocities of about 8–10 m s^{-1} (Romanenko et al. 1973). This must not be very functional since most, if not all, pinnipeds are probably incapable of such high speeds. Those animals that do achieve such speeds are all bare-skinned.

As for the bare-skinned whales, it has been proposed by Kramer (1965) that they have a special skin morphology that causes dampening of differential pressures along the body surface. This stabilizes the boundary layer at high Re. The evidence is based on a hydromechanical model patterned after some observed structures in the whale skin.

11.1.2.3 Surface Secretion

Coating of the skin with drag-reducing compounds was first proposed for fish. It was shown that fish mucous at concentrations as low as 11 ppm was enough to have significant effects on the friction of turbulent water (Rosen and Cornford 1971). However, there was no consistent relationship among fast- and slow-swimming species. More recently, it has been noted that some compounds of cetacean skin have drag-reducing properties (Gucinski and Bauer 1983). How effective some of these chemical and structural modes of drag reduction really are for the animal remains to be determined.

Table 11.3. Swimming speeds of aquatic vertebrates. The records listed are only those measured with velocity meters, or animals tested over a fixed course

Species	Length	Dive or cruise speed		Maximum speed		Reference
	(m)	(m s^{-1})	(Body lg s^{-1})	(m s^{-1})	(Body lg s^{-1})	
Eudyptula minor				1.7		Clark and Bemis (1979)
Spheniscus demersus		1.8		3.2		Nagy et al. (1984); Clark and Bemis (1979)
A. patagonicus		2.4	2.7	3.4	3.8	Adams (1987); Clark and Bemis (1979)
A. forsteri	1.0	2.1	2.1	2.7	2.7	Kooyman et al. (1971 a)
P. caspica	1.3			3.5	2.8	Mordvinov (1968)
P. vitulina	1.6	1.9	1.7	4.9	3.0	Williams and Kooyman (1985)
Leptonychotes weddellii		2.5				Kooyman (1968)
Z. californianus	1.6	2.3	1.4	5.5	3.3	Ponganis and Kooyman (unpubl.)
S. attenuata	1.9			11.0	5.9	Lang (1975)
L. obliquidens	2.1			7.8	3.7	Lang (1975)
T. truncatus	1.9			8.3	4.4	Lang (1975)
Homo sapiens	1.7	1.2[a]	0.7	2.0[b]	1.2	McWhirter (1977)

[a] English channel.
[b] 100-m sprint.

11.2 Behavior

11.2.1 Swim Speed and Size

The swim speeds summarized in Table 11.3 were measured with attached meters, or from animals trained to swim a specific circuit. The cruise speeds of the king penguin, *A. patagonicus*, sea lion, and *Z. californianus*, are the average values obtained from many hours of recording animals foraging at sea. A small, but unspecified amount of drag is induced by the recording device, which may make these values lower than normal.

The top speeds tend to be lower than those reported by observers from ships, a situation in which it is difficult to obtain reliable information. In regard to shipboard observations, it is unfortunate that two reports obtained from ships are so widespread in the literature, in which power requirements have been calculated, because it is likely that they are incorrect, judging from the measured values.

In one of these reports, dolphin speeds were based on animals swimming between two ships, one being towed by the other (Stevens 1950). It was estimated that the speed of a 2-m dolphin was 10.2 m s^{-1}. This was most likely a burst effort in which the animal may have been getting help from bow waves and wake of the ship.

The other often-cited report is that of Johannessen and Harder (1960), in which they did not make the observations, but utilized questionnaires placed aboard various ships. In five reports the dolphins were in large pods, and most approached no closer than 200 m. Again the speeds for 2-m-long dolphins was estimated at about 10 to 11 m s^{-1}, which was sustained for 10 to 25 min. It was concluded that they could sustain speeds of 9.3 m s^{-1}. This is a much higher sustained speed than was found for a trained dolphin, *T. truncatus*, of similar size whose top speed was 8.3 m s^{-1} for 7.5 s, 6.1 m s^{-1} for 50 s and only 3.1 m s^{-1} indefinitely (Lang 1975). Furthermore, it was noted by observers participating in the test of a *Stenella attenuata*, which reached a measured top speed of 11 m s^{-1}, that they all had a subjective impression of speeds at least 25% higher. Finally, the astounding observation that a pod of pilot whales, *Globicephala melaena*, remained with a ship for several days that was traveling at 11.3 m s^{-1}, and that of a killer whale, *Orcinus orca*, approaching a ship at about 15 m s^{-1} (Johannessen and Harder, 1960) need to be verified.

It seems intuitively logical that larger animals, such as the killer whale, should swim faster because drag increases as the surface area or L^2, whereas power increases according to muscle volume, or L^3. According to Aleyev (1977), this relationship is true for L ≤ 4.5 m, about the length of pilot whales or killer whales, but reverses at greater lengths. Aleyev argues further that swim speed is also directly proportional to frequency of locomotion movement. However, maximum frequency is inversely related to muscle length; thus the speed decreases. As an example, the maximum tail beat frequency of a small dolphin is about 10 hertz; that of a large whale is about 2 hertz.

It should be noted also that characteristics of flow change from small to large bodies. At the higher Re, at which large whales must swim, C_D becomes constant rather than continuing to decline (Fig. 11.2). Thus, this advantage of size is also lost in exceptionally big animals. These physical features of streamlined animals make them a prisoner of their profiles: What other strategies can be employed to reduce drag other than the physical adaptations mentioned above?

11.2.2 Burst and Glide

In theory, there are also some behavioral means of reducing drag. One of the most familiar is the brief swim phase followed by coasting. Burst and glide theory has been reviewed in detail by Blake (1983 b), whose discussion centers around the original proposal of Weihs (1974). The argument is based on the premise that over a given distance the cost will be less than if steady swimming prevails. From Fig. 11.6, it can be seen that the lower the average velocity, the greater the benefit of burst and glide. It was reliance on this behavior that has enabled various investigators to obtain the drag coefficients presented in Table 11.2. That this method of swimming is actually used is clear from observations of both wild and captive seals, sea lions and penguins. I have been impressed by the fact that at preferred swim velocities in pools and in the wild much of swim time is spent gliding.

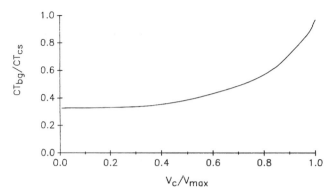

Fig. 11.6. Ratio of energy per unit distance traveled in burst and glide swimming (CT_{bg}) to that for constant swimming (CT_{cs}) plotted against normalized velocity. Vc is average speed; V_{max} is maximum sustainable speed. (After Weihs 1974)

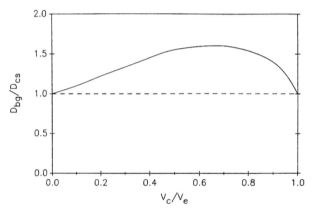

Fig. 11.7. Range (or endurance) of burst and glide mode of swimming relative to steady swimming is plotted against normalized velocity for salmon. D_{bg} is distance traveled in burst and glide swimming and D_{cs} is distance traveled in constant swimming. V_c is average speed and V_e is maximum sustainable speed. (After Weihs 1974)

This type of swim behavior can result theoretically in energy savings of up to about 80%, according to Weihs's original equations (Fig. 11.7). However, Blake (1983 b) summarizes various modifying conditions that would reduce this savings. Insufficient augmentation of drag for the high phase of burst and glide, over that of steady swimming, was incorporated into the original estimates. Also, Weihs assumed equal propulsive efficiency for both steady swimming and burst and glide, whereas steady swimming may be greater. When these are incorporated, the savings are reduced to about 50%. Finally, the theory was developed for caudal propellors; how it applies to pectoral thrusters such as sea lions and penguins remains to be proposed.

138

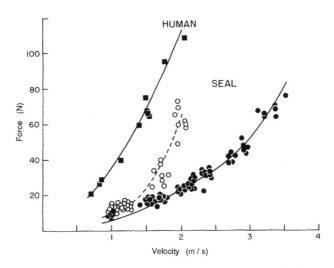

Fig. 11.8. Augmentation of the drag coefficient (C_D) due to surface interference is plotted relative to the ratio of depth of submersion over maximum body depth. (After Blake 1983 b)

Fig. 11.9. Total drag (N = Newtons) in relation to velocity of a towed human and trained seal of approximately the same body mass. *Open* and *closed symbols* are surface and submerged tows. (After Williams and Kooyman 1985)

11.2.3 Porpoising

Another hypothesis in which behavior plays an important role in energy savings is porpoising. This proposition applies to any animal that swims near the surface. The basis of the idea comes from the observation of Hertel (1969) that when a streamlined body of FR = 5.6 is towed near the surface, C_D increases markedly as the depth decreases from about 3 to 0.5 maximum body diameters (Fig. 11.8). This increase of up to five times was verified in seals towed at 2 m s^{-1}, in which the surface drag was three times that when submerged at 2.8 body diameters (Fig. 11.9). The factor appears to become even greater at higher speeds, but the seals refused to be towed on the surface at speeds above 2 m s^{-1}.

Noting these large increases in drag when a streamlined body is near the surface, Au and Weihs (1980) developed an expression of the energy savings of por-

139

poising. This was based on the comparison between an animal swimming continuously near the surface, and the energy saved over the distance of the leap. For this reason, they estimated that the animal would leap at a 45° angle to cover the greatest distance. However, Gordon (1980) points out that a 45° angle of departure would result in a loss of speed, and it is more likely that the departure angle would be 30°. This shallower angle of leaping has been observed in porpoising penguins (Hui 1986).

In addition, no drag augmentation factor was included for oscillatory swim movements. Blake (1983 a) revised the equation with this value equal to 4. This results in a lower estimate of critical velocity to porpoise from that of Au and Weihs. A 2-m animal should porpoise at about 2.5 to 4 m s⁻¹ depending on a fully turbulent, or 50% turbulent boundary layer.

These estimates do not take into account the respiration rate, duration of time necessary for a breath, or the sea conditions, to mention a few. Au and Weihs (1980) used in their argument that this savings would apply when ventilation rate is so high that it must be continuously near the surface. Such a high rate, however, probably never occurs in nature. Rather, the energy savings is more a question of whether it is advantageous in some way for the animal to completely pass through the air/water interface in a sinusoidal path that only briefly is <3 body diameters deep, or to project the nose only above the surface. This probably varies among species; it should be noted that there are at least two high-speed swimmers of about 1.5 to 2 m length that do not porpoise. These are Dall's (pers. commun.) and harbor porpoises (D.A. Croll, pers. commun.).

11.3 Power Requirements

From the basic drag equation, the power requirements of the animal can be estimated if a few assumptions are made. One is that for a streamlined body of FR $\sim$ 5 pressure drag is about 20% of the frictional drag, so total drag can be estimated as 1.2 C_F [Eq. (11.6)]. Also, the overall aerobic efficiency of about 25%, estimated by Webb (1975) for small salmon (0.65 m), is used. The equation is expressed as:

$$P = \frac{1/2 \varrho A V^3 C_D 1.2}{0.25},$$
(11.7)

where $C_D = 1.3 R^{-0.5}$ in laminar flow and $C_D = 0.072 R^{-0.2}$ in turbulent flow. This does not account for an augmented drag due to swim motion, which was assumed earlier, to increase drag and thus power requirements by three to four times. For a harbor seal I have assumed a 30% laminar flow for reasons discussed earlier on glide C_D. Then a 1.7-m seal swimming at 1.4 m s⁻¹, near its preferred velocity, would have an aerobic power requirement of 13 W at 100% laminarity, 45 W if 100% turbulent, and 32 W if 30% laminar. The measured MR of a seal swimming at 1.4 m s⁻¹ is 208 W (Davis et al. 1985). Since the resting MR was 113 W, that would leave 95 W for the power required to swim at

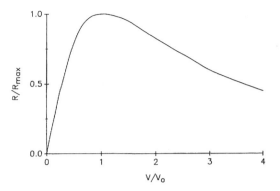

Fig. 11.10. Ratio of range (R) to maximum range (R_{max}) relative to normalized velocity for constant speed swimming where V = cruising speed and V_0 = optimal cruising speed. (Weihs 1973)

1.4 m s^{-1}. That is three times higher than the calculated need, and appears to fit the three to four times drag augmentation factor with 30% laminar flow. However, the water flow in the flume was probably turbulent so that the higher estimate of 45 W is probably more realistic for these conditions. Thus, drag augmentation would be no more than two times.

These results are interesting in light of the theoretical derivation of optimum swim speed for fish in terms of maximum range per total energy store (Weihs 1973). According to theory, the optimum cruising velocity (V_m) is that value in which the energy expenditure rate is equal to twice the resting metabolism of the animal. The decay in achievable range is rather slow at higher speeds where at four times the V_m the range is halved (Fig. 11.10). The calculated level of metabolism of twice resting is about that which was determined for a seal swimming at 1.4 m s^{-1} in turbulent flow (Davis et al. 1985). Presumably the speed of the seal at the lowest cost of transport would be higher in still water. Also, the average daily metabolic rates of foraging fur seals (Costa and Gentry 1986), and the MR of a dive bout in Weddell seals are of similar magnitude of about two times the resting MR (Kooyman et al. 1973). These seem close to the prediction by Weihs.

Such estimates can be applied to the O$_2$ store, as well as to the stored carbon energy reserves to which Weihs was referring. It is the rationale for the O$_2$ utilization rate, intrinsic to the drag and aerobic power requirements, that are important for an estimate of the aerobic dive limit, the subject of the following chapter.

11.4 Conclusions and Summary

In this chapter I have dealt with various aspects of the problem of movement in water, in preparation for Chapters 12 and 13. The basic question in Chapter 11 has been: Why animals swim at the speeds they do? To resolve this question some parts of basic hydrodynamic theory have been discussed. Some of the most important considerations in hydrodynamics are the influences of viscosity and density on drag, which consists of form, friction, and induced drag. Equations of motion show the influence of Re and C_D on the drag force. A powerful influence is speed to which drag is a function of V^2 Most cruising speeds of aquatic animals up to about 3 m in length appear to be in a range of 1 to 2 m s^{-1}, which keeps drag low and at a rate that appears to be near the minimum cost of transport. Drag is also reduced by burst and glide swimming, which is probably common at these low speeds. How shape and surface properties of the animal, as well as behavior, influence power requirements is speculative. Theoretical arguments are available, but little experimental data.

Chapter 12

Aerobic Dive Limit

It seems to me that one of the most important objectives of diving physiology would be the ability to predict the limit of duration of aerobic dives by aquatic tetrapods. Such predictability would be a valuable asset in understanding the behavior of the animals and their ecology. In addition, such a prediction capacity would require a considerable understanding of the physiology of the animals, particularly partitioning of O_2 stores and their management. The following chapter with its crude estimates indicates in a broad way the strengths and weaknesses of our present information on diving physiology and management of O_2 stores.

12.1 Definitions

Aerobic dive limit (ADL), as I defined it in an earlier report, is "the maximum breath-hold that is possible without any increase in blood LA concentration during or after the dive. This limit is dependent upon available O_2 stores, oxygen consumption rate ($\dot{M}_{O_2}$), degree of peripheral vasoconstriction, and rate of LA production and consumption". It has been called an "aerobic dive limit" rather than a "lactate threshold limit" to emphasize two metabolic characteristics of such dives: (1) They are fueled primarily by aerobic metabolic pathways; and (2) There is no net increase in anaerobic metabolic products above steady-state resting or surface swimming metabolism. I have used aerobic dive limit in preference to aerobic limit because the latter is ambiguous to me and could mean either the maximum breath-hold capacity of a resting animal before obligate aerobic organs dysfunction, or the maximum breath-hold capacity of a resting animal before any increase in blood lactate levels occurs. By this definition the aerobic limit (AL) should be longer than the ADL since the latter refers to an animal actively swimming and searching for prey. Presumably animals do this at consistent speeds and therefore the ADL is predictable if the speed is known. As will become obvious, this is crucial to the calculation because $\dot{M}_{O_2}$ is directly related to the swim rate (Chaps. 10 and 11).

Finally, the choice of ADL first used by my colleagues and I (Kooyman et al. 1983) instead of aerobic dive time (ADT) used by Hochachka and Somero (1984) is minor. I have used "limit" to imply that the value reflects some kind of a boundary that cannot be exceeded without alterations in O_2 management responses, whereas to me this is not implied in ADT.

As will become obvious, my calculations of ADL are based on total available O_2 store of Table 5.2. Consonant with the earlier statement, a diver cannot ex-

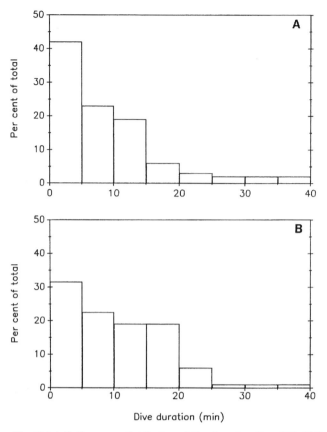

Fig. 12.1 A, B. Frequency distribution of dive durations for: *A* Weddell seals diving from an offshore, isolated breathing hole that was cut in fast ice. The ice hole was in territory unfamilar to the seals. Total of dives was 1057. Maximum dive duration was 73 min. *B* Summary of 22 free-ranging seals. Total of dives was 4601. (After Kooyman et al. 1980)

ceed the submersion limit which uses the entire store, without some tolerance to anoxia in *all* organs. Such a tolerance is not known to exist in any vertebrate except some freshwater turtles, and in those instances they are not diving, but experiencing "resting" submersion and torpor.

For an animal to exceed its ADL, some organs must become completely or partially anoxic as blood flow is restricted, and may then rely on hypometabolism and/or anaerobic metabolism. The model presented earlier (Fig. 8.10) accounts for skeletal muscle that becomes anaerobic. Other organs such as the kidney and gastrointestinal tract may become hypometabolic as function declines. Because dives in excess of the ADL are anticipated, flow restriction occurs early, as noted for the kidney (Fig. 7.2). When the internal O_2 store for an organ is depleted, LA accumulates rapidly, hence the exponential increase in its production as dive duration increases (Fig. 8.4). However, based on blood [LA] and dive durations of Weddell seals, the ADL exceeded $<5\%$ of the time in adults (Fig. 12.1), and sub-adults (Fig. 12.2). Consequently, the reason for this

144

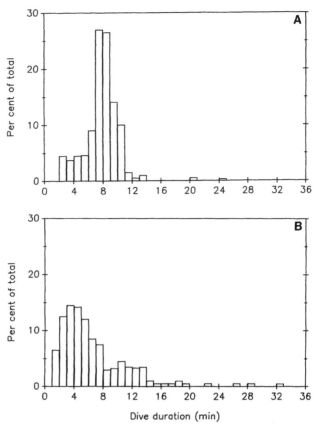

Fig. 12.2 A, B. Frequency distribution of dive durations for 2 seals with average mass of 140 kg (*A*) and 1 seal of mass 200 kg (*B*). Total of dives was 264 for *A* and 370 for *B*. (After Kooyman et al. 1983)

chapter is to discuss the possibility of calculating this important variable, which can be verified directly by measurement of blood [LA] in only a few special cases.

12.2 Calculations

Estimates of ADL do have some risks of embarrassment because of so many un-evaluated assumptions, which may cause large errors. Confidence in calculation of ADL stems from the close correlation, as noted below, of observed and calculated values for the Weddell seal. The ADL does not tell us how the system works; I have attempted that in Chapters 7 and 8. The ADL only gives us an end-result, and its probable influences on the behavior of the animal.

A demonstration of how absurd the estimates can become, if one is not careful, is a humorous tongue-in-cheek calculation that I recall Dr. D. R. Jones making after a serious attempt to scale and estimate the maximum submersion tolerance before asphyxiation in the domestic mallard (Hudson and Jones 1986). Testing a series of variables, they derived an equation that matched their observations for ducks ranging in size from 0.05 to 3.5 kg, and found that the forced submersion limit relationship was:

$$SL = 6.6 \, Mb^{0.64}, \tag{12.1}$$

where Mb is in kg and SL in min. If the same scaling factors were applied to larger animals, such as a mallard the size of a Weddell seal, it could remain submerged about 5.5 h, or I reckon 4.4 days for a mallard as large as a sperm whale! This does not tell us much about larger animals, but it does point out both the hazards of violating statistical rules and extrapolating regression equations beyond their data base, and perhaps even how remarkable the domestic mallard is. I once called the Weddell seal a "consummate diver" (Kooyman 1981), perhaps the title is more fitting to the mallard as a consummate subject for forced submersion.

On the more serious side, the two major variables on which ADL is dependent are the O_2 stores and metabolic rate. The latter is directly coupled to swim velocity. How critical swim velocity is was discussed in Chapter 11, where it was shown that metabolic rate varies with the cube of velocity.

The influence of body size on ADL in the same species has been shown for Weddell seals (Figs. 12.1 and 12.2). These determinations are discussed below in relation to standard calculations of ADL. Calculations for selected marine reptiles, birds, and other marine mammals are discussed also. Some species were selected because frequency distributions of dive duration have been determined (Chap. 13).

The ADL of several species, and those for the same species done by different investigators, are reported in Table 12.1. Those estimated by me use the available O_2 stores as summarized in Table 5.2. $\dot{M}_{O_2}$ was estimated as some multiple of SMR using the equations of Chapter 10. The other estimates of ADL are discussed in comparison.

12.2.1 Reptiles

Reptiles present the largest discrepancy of calculations of ADL and of observed breath-hold patterns. Most dive durations are well below those estimated. For example, captive loggerhead sea turtles in a pool remained submerged on average 5 min, but their estimated ADL is 33 min (Lutz and Bentley 1985). This in part may be a difference between resting submergence and hunting dives when blood flow distribution may not be the same. The latter follows the model of Fig. 8.14. In the resting state flow may broadly perfuse muscle and viscera continuously, resulting in a more rapid depletion of blood O_2 than if it is more restricted and pulsatile, as I have proposed for hunting dives. Also, the surface is so conveniently available in a shallow pool that there is little incentive for extended breathholds.

146

Using the same available O_2 store of the loggerhead turtle for the leatherback sea turtle, and extrapolating the $\dot{M}_{O_2}$ of Eq. [1] of Chapter 10 to a 450-kg turtle, a dangerous statistical violation as I pointed out earlier, I also obtain an ADL much greater than the observed dive durations. The calculated ADL is 44 min. Most of the dives measured were 12 to 15 min; <1% exceeded about 20 min (Fig. 13.2). I suspect that the dive results more closely estimate the ADL. In fact, if for reasons expressed in Chapter 4 regarding the liability of the lung as a N_2 store as well as an O_2 store during deep diving we take the extreme case that no O_2 from the lung is available, then the ADL would be about 12 min.

12.2.2 Birds

Little is known about dive durations in birds outside of enclosed pools to compare with the calculated ADL for penguins. In most cases the dive durations noted are <1 min, but emperor penguins commonly dive for 2 to 8 min (Kooyman et al. 1971 a, Fig. 13.6). Since on the occasion of most of these dives (Fig. 13.5) the birds appeared to be hunting, neither the ADL estimate of 4.1 min nor 2.3 min seems realistic (Table 12.1), otherwise most of the observed dives in Fig. 13.5 were anaerobic. The LA accumulation over several hours of this kind of diving would presumably be intolerable.

The lack of agreement on ADL for emperor penguins between Butler and Jones (1982) and Kooyman and Davis (1987) in view of these recent observations seems almost academic, but it was for the following reasons: (1) we included the muscle O_2 store, which would add about 25% to the total O_2 store, and (2) we as

Table 12.1. Calculated aerobic dive limit. See Table 5.2 for a summary of O_2 stores and Tables 10.1 and 10.2 for a summary of metabolic rates

Species	Mass	Swim velocity	Swim $\dot{M}_{O_2}$	ADL	Reference
	(kg)	m s^{-1}	Resting $\dot{M}_{O_2}$	(min)	
Caretta caretta	20	0	1	33	Lutz and Bentley (1985)
Dermochelys coriacea	400	0.5	2	44	This chapter
Eudyptula minor	1.2	0.7	2	1.7	Kooyman and Davis (1987)
Spheniscus humboldti	4.5	?	1.5[a]	2.3	Butler and Woakes (1984)
Spheniscus humboldti	3.8	1.0	2	2.5	Kooyman and Davis (1987)
Aptenodytes forsteri	24	?	3	2.3	Butler and Jones (1982)
Aptenodytes forsteri	25	2	2	4.1	Kooyman and Davis (1987)
Fur seal, any species of this mass	40	2	2.5	3.5	Gentry et al. (1986 a)
Zalophus californianus	90	?	2	5.6	Feldkamp (1985)
Leptonychotes weddellii	140	1–2	1.6	12	Kooyman et al. (1983)
Leptonychotes weddellii	200	1–2	1.7	13	Kooyman et al. (1983)
Leptonychotes weddellii	450	1–2	1.6	19	Kooyman et al. (1980)
Leptonychotes weddellii	450	?	1.6	21	Hochachka and Somero (1984)
Leptonychotes weddellii	450	0.23	0.45	65	Guppy et al. (1986)

[a] Measured resting value was 1.5 × predicted from Eq. (3), Chapter 10.

sumed a lower $\dot{M}_{O_2}$ during hunting dives (Table 12.1). Taking these variables into account reduces the difference between the two estimates.

The ADL's estimated for the Humboldt penguins were similar at 2.3 and 2.5 min (Butler and Woakes 1984; Kooyman and Davis 1987), but for different reasons. We used an $\dot{M}_{O_2}$ of two times SMR whereas theirs was three times. Also, they did not include the muscle O_2 store, which could add about 25% to body O_2 store. Finally, the resting $\dot{M}_{O_2}$ used by Butler and Jones is 1.5 times that predicted [Eq. (10.3)], which puts it at a level close to that used by Kooyman and Davis (1987) for their calculations of the diving bird. Taking these differences into account, the ADL for the two reports are equivalent.

The level of $\dot{M}_{O_2}$ above SMR is arbitrary at this time because so few measurements have been done. Based on time-partitioning studies, Nagy et al. (1984) estimated the dive $\dot{M}_{O_2}$ of jackass penguins as nearly ten times SMR. If this estimate is used, then ADL of similar sized Humboldt penguins would be 0.4 min! The estimate seems unrealistic for three reasons: (1) Such a low ADL gives little time for diving to depth and searching for food. (2) Measurements of $\dot{M}_{O_2}$ of little blue (Baudinette and Gill 1985) and Humboldt penguins (Hui, 1983) reached only two and four times resting $\dot{M}_{O_2}$ at about expected swim velocities for diving (Table 10.1). It should be noted also that in the studies of both the little blue and Humboldt penguin the birds were probably swimming in turbulent conditions in which swim effort was greater than would exist in natural conditions. Finally, (3) Nagy et al. (1984) commented that such a high level was not surprising because it was commensurate with the cost of flying birds. However, within the same report they noted that in flying swifts and sooty terns cost of flying was only 2.9–5.3, and 4.8 times SMR, respectively, so it would seem that the argument is not consistent.

12.2.3 Mammals

Somewhat more is known about $\dot{M}_{O_2}$ of swimming seals. Several estimates of ADL of Weddell seals have been made using O_2 store values of Table 5.2 and $\dot{M}_{O_2}$ levels of Table 12.1. The smallest seals have the lowest calculated ADL. This is because the $\dot{M}_{O_2}$ has been scaled to body mass to the exponent of 0.75, but available O_2 store is constant for any size. However, the measured ADL is about 6 min in the smallest seals, or about half the 12-min prediction (Fig. 8.4). Behavioral observations suggest that the smaller seals may have been swimming proportionally faster to reach the same depth to feed as the adults, and possibly to capture prey. This would raise $\dot{M}_{O_2}$ proportionally higher in the smaller seal.

ADL for adult Weddell seals have been calculated to be 19, 21, and 65 min. The 19 and 21 min estimates are not as close as seems apparent. Using the same figure of 1000 mM of blood O_2 and a $\dot{M}_{O_2} = 74$ m$\dot{M}_{O_2}$ min^{-1} (Hochachka and Somero 1984) I calculate 13.5 min rather than 21 min. Also, blood O_2 extraction cannot be complete, and if available O_2 is defined as in Table 5.2, then only 750 mM blood O_2 is available, which makes ADL = 10 min. However, this $\dot{M}_{O_2}$ is based on whole body metabolism (Kooyman et al. 1973 a), therefore, muscle O_2 store should be included. Based on an estimate of 500 mM of muscle

O_2 (Hochachka and Somero 1984) + 750 mM blood O_2, the ADL would be 17 min. This is in accord with estimates of Kooyman et al. (1980), but the muscle O_2 estimate is higher in the former. Based on variables in Table 5.2, muscle O_2 would equal 2.7 mM kg^{-1} or 360 mM O_2 in 135 kg muscle of a 450-kg seal, rather than 500 mM.

The estimate of 65 min is based on an $\dot{M}_{O_2}$ of 0.45 SMR and swim velocity of 0.23 m s^{-1} (Guppy et al. 1986). None of these assumptions is supported by existing data on swim velocity (Kooyman 1968), $\dot{M}_{O_2}$ (Kooyman et al. 1973 a; Davis et al. 1985), or blood [LA] (Kooyman et al. 1980; Kooyman et al. 1983; Guppy et al. 1986; Qvist et al. 1986). Therefore, I believe it is very unlikely that Weddell seals can remain overall aerobic for a 65-min breath-hold under any natural conditions. However, it should be noted that if SMR is reduced about 30%, such as when the seal is sleeping, this would make possible completely aerobic dives in northern elephant seals even when dive durations are about 40 min. It would not be necessary in that case to invoke the highly speculative curve illustrated in Fig. 8.8 if during such dives the seal sleeps.

12.3 Conclusions and Summary

A definition of ADL is given and calculations of ADL are made in accordance with this definition. Some differences in results are discussed and it is concluded that for mammals the ADL estimates are close to actual observations of either blood lactate increases, or the dive durations to which the animals limit themselves.

The ADL tells little about how O_2 stores are managed, or the variation in physiological responses necessary to achieve different breath-hold limits. However, it focuses attention on some important areas of aerobic breath-hold diving that need investigation. For example, the Mb O_2 store must be important, and its management a critical function. Also, the ADL calculations indicate the importance of metabolic rate, which is directly related to swim velocity. This raises questions about hunting strategies. Some animals may swim fast, resulting in a short ADL. What they gain is covering a greater distance in a shorter time in search of prey. Others may swim more slowly, perhaps at the speed of minimum cost of transport to gain a maximum submersion time and the greatest distance possible, but at a slower swim rate.

Chapter 13

Behavior

This could have been either the first or, as it is, the last topical chapter. If it had been at the beginning the reader would have had a point of reference for the physiology discussed. Instead I chose it to be the culmination of all, to which previous chapters have been directed. That is to answer the question of how the internal workings of the animal make possible the way in which these divers hunt and make a living at sea.

In Chapter 4 a table and discussion on maximum depths was presented to impress the reader with the pressures that divers may endure. As remarkable as these maximum depth capabilities are, of equal or greater interest to comparative physiologists are the physiological adaptations to routine foraging behavior, which has been the major objective of this book.

In Chapter 10 I discussed levels of metabolism, and then later how they are influenced by swim velocity (Chap. 11). These are key issues in dive behavior and understanding of hunting strategies, i.e., how the diver uses its swim velocity and metabolic rate to catch prey.

These variables are the cornerstone to deciphering the way in which O_2 stores are partitioned (Chap. 5), variation in stores throughout the dive (blood and muscle gases – Chaps. 8 and 9), and how they are managed. The management of the store is the crucial factor in determining the aerobic dive limit, a value that was discussed extensively in Chapter 12, and to which reference will be made again in this chapter as I review the dive duration frequency distribution of species. In fact, the percent dive time becomes a matter of special interest in the following comparison of species because there is a dichotomy in these diverse divers.

The work on marine divers has shown that they might be divided behaviorally into two broad types, "divers" and "surfacers." The differentiation, which was first expressed by Kramer (1988), a behavioral ecologist, appears to me to be an appropriate and important classification. Divers are those animals that over a 24-h period at sea usually spend most of their time, i.e., more than 50%, on the surface. These dive periods are often circadian and usually occur in bouts that may last several hours as the animal hunts and captures prey. Bouts are best described as a series of dives, usually to a consistent depth, with a short surface interval between each dive for gas exchange. During the bout, most of the time, about 80 to 90%, is spent under water. The duration of the dives is most likely dictated by O_2 stores and O_2 consumption rates. Thus the fur seal with its smaller O_2 store relative to its high metabolic rate makes short dives compared to the seal. During the bout the animals might be considered surfacers rather than divers, but this term seems more appropriate to another definition.

Surfacers are those animals that, based on present data, spend most of their time at sea under water and surface only briefly for unloading and loading of CO_2 and O_2, respectively. Diving appears to be continuous. As new data are accumulated, they may reveal that breaks in this pattern may occur only seasonally for breeding or molting when the animals must be ashore. So far there are three examples, two reptiles and a mammal, each of which is discussed below. I suspect that as the dive behavior of more species are studied more surfacers will be discovered.

For the rest of the chapter I will follow the order of previous chapters by discussing reptiles first, followed by birds and mammals. Much of the information is new and has not been published as of this writing because of the rapid accumulation of data due to recent developments in recording systems. Some of the terms may be unfamiliar to physiologists, such as dive bout, which I have explained above, bottom time, patterning, and deep scatter layer.

In brief, bottom time is the period spent at depth exclusive of the descent and ascent time. Patterning of dives becomes apparent in the time of day or night that dives occur, the rate of diving, and the consistency of the dive depths. In some of these the patterns become machine-like in the consistency of the variables mentioned above. Some of the uniformity in depth is thought to be due to prey distribution which, in pelagic species, is part of the deep scatter layer community. This is a midwater area of animal concentration that, due to the acoustic reflective properties of some animals, causes depth sonar equipment to indicate a false bottom (Dietz 1962). It is an area of high animal concentration in which many of the species migrate vertically on a diurnal cycle. It is also found in all oceans of the world and may be the most broadly distributed, and largest in area and number of animals of any biological community on this planet.

13.1 Reptiles

So far, the diving habits of three species of reptiles, the marine iguana, the yellow-bellied sea snake, and the leatherback sea turtle have been studied.

13.1.1 Marine Iguana

Of the three species, the marine iguana is an exceptional case, frequenting only the intertidal and subtidal zones where as a strict vegetarian it forages on marine algae. The species divides itself into intertidal and subtidal foragers by size. Those < 1200 g feed intertidally and those > 1200 g feed at sea (Trillmich and Trillmich 1986). Of those that feed at sea only 5% of the day is spent close to or in the water. This limited use of the sea as well as other characteristics of the species has prompted the argument that this is a land lizard with no aquatic specialization except its food habits (Dawson et al. 1977). Such a land-bound existence is greatly different from the other two species of marine reptile.

13.1.2 Yellow-Bellied Sea Snake

The yellow-bellied sea snake is pelagic, seldom comes ashore, bears live young, and in a recent tracking study was shown to spend 87% of its time underwater (Fig. 13.1; Rubinoff et al. 1986). This species would be considered a "surfacer" or continuous diver by the definition given above.

About 65% of the dive profiles have four phases (Fig. 13.1) (Graham et al. 1987). There is a rapid descent, bounce ascent, gradual ascent (82% of the dive), and a final ascent. Some of these dives are very long, one was nearly 4 h. Of special interest is that most are beyond the predicted O_2 reserves or ADL. The ADL is confounded because there is cutaneous gas exchange in this species while diving (Graham 1974). To explain these long dives it was suggested that during much of the time aerobic metabolism was reduced or cutaneous respiration was increased, or both (Rubinoff et al. 1986). If the snakes were merely resting at depth this explanation may be adequate, but not if there is much active swimming, unless cutaneous respiration is much greater than previously measured. Furthermore, on the basis of the depth and duration of the dives, two internal responses seem necessary for gas exchange to be effective in management of gas stores: (1) There must be an intracardiac shunt to meter lung O_2 depletion and enhance cutaneous exchange without loss of O_2 stores to sea water. (2) Enough cutaneous exchange should occur to promote loss of N_2 to sea water rather than distribution to body tissues when N_2 partial pressure is increased at depth due to lung compression. (See Chap. 4 for a discussion of deep diving and decompression sickness).

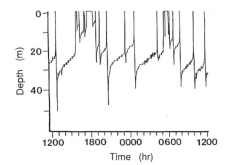

Fig. 13.1. Dive profiles of a yellow-bellied sea snake. (After Graham et al. 1987)

13.1.3 Leatherback Sea Turtle

Because the sea turtle also has a large lung volume, a similar intracardiac shunt to meter lung gases to the body may be necessary to adapt to the dive schedule. This species is the only one in which a detailed study of diving has been done. It is the largest and most pelagic of all turtles. At present there is one published account on two turtles (Eckert et al. 1986), and another in preparation on seven others. All were females (350–450 kg) diving during the internest interval.

Distinct from the sea snake the dive durations are not exceptionally long with a mean of 11 min (Fig. 13.2). They are shorter than a seal of comparable size (see

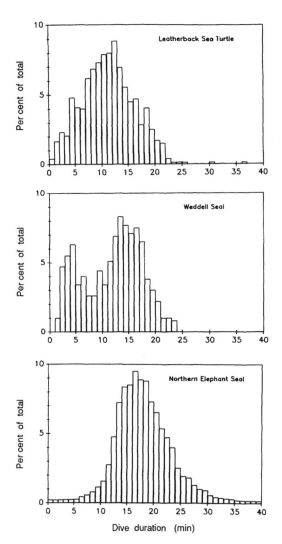

Fig. 13.2. Frequency of occurrence of dive durations from two 10-day records of female leatherback sea turtles. Total of dives was 1269 (data from Eckert et al. 1986). Frequency of occurrence of dive depths of Weddell seals during December. Total of dives was 505. (M.A Castellini, R.W. Davis and G.L. Kooyman 1988) Frequency of occurrence of dive duration in female northern elephant seals. Total of dives was 8497 (data from Le Boeuf et al. 1988)

below). Based on this histogram the behavioral or functional ADL, assuming that only 10% of dives exceed this limit, is about 20 min. This ADL is identical to the Weddell seal (see Fig. 12.1), and it is much less than the predicted ADL of 44 min (Table 12.1) Indeed, of the nine turtles studied, the longest dive recorded was 37 min.

Short dive durations are the only example of a species diving for much less than the predicted ADL. Several variables have not been measured as yet in the leatherback, but even large errors in assumptions would not account for this discrepancy. Either the turtle is simply not inclined to dive near the limit of its O_2 stores, or there is a major error in estimates.

Two possible errors that are interrelated are the swim velocity and metabolic rate. Perhaps the MR of the leatherback is much higher than estimated from

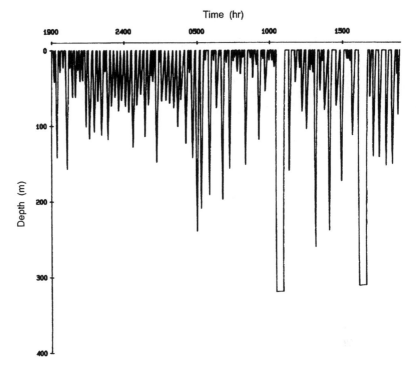

Fig. 13.3. Dive profiles for 24 h of a female leatherback sea turtle. The first offscale dive lasted 35 min, the second 37 min. The depth of both were estimated to exceed 1000 m. (S. Eckert et al. 1989)

regression equations of other reptiles [Eq. (1), Chap. 10], due to their proposed endothermy (Frair et al. 1972; Mrosovosky and Pritchard 1971).

This species is another example of a continuous diver. During the 10-day internesting period only about 10% of the time is spent at the surface. The average surface time is 5 min; the maximum measured was 167 min (Eckert et al. 1986). An example of the continuous dive pattern over a 24-h period is displayed in Fig. 13.3.

This particular turtle and time were selected to show extremes of some interesting and unexplained phenomena. All turtles show a diurnal rhythm in which dives at night are shallower than those during the day. Presumably at night they are feeding on some member of the DSL. More puzzling are the very deep dives about mid-morning and afternoon. In the case illustrated, which includes the most extreme examples on record, the deep dives were unusually long (35 and 37 min) and deep. They exceeded the range of the recorder, but not the limit of the transducer. The tests suggest that these were dives in excess of 1000 m. Such behavior raises baffling questions in regard to the reason for the dives, and their physiological effects. Equally bewildering behavioral observations have been obtained recently from birds.

155

13.2 Birds

Investigations of diving behavior of several species of birds have dealt with the maximum depth of an entire foraging trip. Some of these reports were summarized in Table 4.1; a more comprehensive survey was recently published (Kooyman and Davis 1987). Some of the dive depths observed are impressive in regard to the size of the birds, i.e., an 180-m dive for a 1-kg common murre (Piatt and Nettleship 1985). Although these maximum depths give an impression of the birds' capability, they give little information on preferred and presumably normal feeding depths. Most analyses of the frequency of occurrence of different depth of dives have been done on penguins, which will be the center of the remainder of the discussion on birds, and wherein the physiological mystery exists.

13.2.1 Penguins

The most detailed studies on frequency of occurrence of different depth of dives have been on gentoo and macaroni penguins (Croxall et al. 1988). It was found that the gentoo fed mainly on fish, but did take krill also; the depth of dives was generally between 25 and 100 m. In contrast, the macaroni, which fed exclusively on krill, usually dived to <20 m with occasional dives as deep as 60 m (compare this to the antarctic fur seal, discussed later, which also feeds exclusively on krill; see Fig. 13.12).

For these smaller penguins there is nothing particularly exceptional about the dive depths. This is true of the few durations noted as well for the gentoo, which, as determined by radio telemetry, averaged 128 s (Trivelpiece et al. 1986). The average durations were 91 s for the smaller chinstrap penguin. Such results are no cause for concern to our present physiological explanations of dive durations as discussed in previous chapters. Difficulties do arise, however, when the results of the larger penguins are considered.

Most of the work on king and emperor penguins has been done by my colleagues and I and has not been published. The most recent work evolved from earlier studies in which the maximum depths of 240 m for the king and 265 m for the emperor had been measured (see Chap. 4; Kooyman et al. 1971 a; Kooyman et al. 1982). Depth-frequency analysis for the king penguin showed that dives to >240 m were rare, but that dives to >100 m were common. The previous study of emperors had only a few depth measurements so there was uncertainty about the frequency of deep dives. More recent experiments reveal that deep dives near 200 m are common (Fig. 13.4). For the one bird for which data are illustrated, about 30% of the dives are >100 m, and if the 0–25-m dives are not considered, because many if not most have functions other than feeding, then these deep dives constitute a high proportion of the dives made during a feeding trip. That the birds will consistently work at these depths is illustrated in Fig. 13.5. This figure shows most of a dive bout which I would interpret as a major feeding effort on the part of the bird. The bout lasted for 4 h, during which much of the time was spent at depths of 150 m or more.

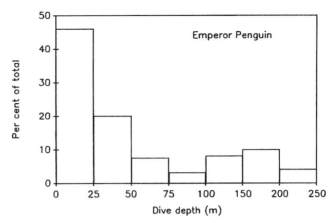

Fig. 13.4. Frequency of occurrence of dive depths from a 6-day record of an emperor penguin. Bars are labeled with the deepest dive, i.e., 25 = 0–25 m. (Kooyman unpubl.)

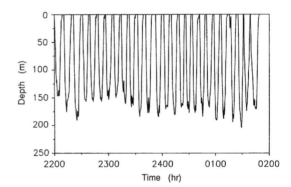

Fig. 13.5. Dive profiles of an emperor penguin during a dive bout. (Kooyman unpubl.)

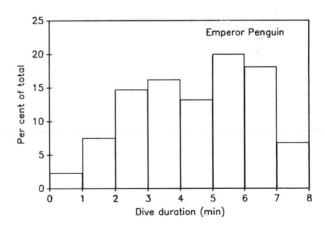

Fig. 13.6. Frequency of occurrence of dive durations from a 6-day record of an emperor penguin. (Kooyman unpubl.)

157

These observations raise two problems that are difficult to explain. First, there are long exposures to high levels of compression. Because each dive lasts 6 to 8 min, the hypotheses discussed in Chapter 4, primarily short exposure and low pressure, for avoidance of bends by birds, do not fit the observations. Second, the gas volumes with which birds dive (Chap. 4) are large. This means, unlike mammals, a substantial gas volume for N_2 absorption is available on each dive. Any proposed reduction in this volume reduces an important O_2 store (Table 5, Fig. 5.3) and heightens the difficulties of explaining the inconsistencies in aerobic dive limits.

It is presumed that through a dive series such as illustrated in Fig. 13.5 the birds remain aerobic for the reasons discussed in Chapters 8 and 10. When the ADL is calculated (Table 12.1) we find that even when including a large respiratory gas volume, based on a full inspiration before the dive, it falls short of observations. A frequency distribution of dive durations (Fig. 13.6) also shows a high proportion of dives which exceed the calculated ADL of 4.1 min.

In brief, it seems that behavioral studies have shown that we do not understand the physiology of diving in birds that hunt at depths > 100 m. In contrast, the following discussion on the behavior of mammals fits reasonably well with our understanding of diving physiology, at least in phocids and otariids.

13.3 Mammals

The most comprehensive investigations of dive behavior have been done on marine mammals – namely phocids and otariids. Behavioral studies in which some type of attached depth recorder was employed have been conducted since at least the 1930's, when Scholander (1940) placed manometric type maximum depth recorders on harpoons shortly before they were imbedded in fin whales. A more detailed study of Weddell seal was done using similar maximum depth recorders, as well as time/depth recorders (Kooyman 1968). Most recently a broad comparative study of fur seals has been carried out (Gentry and Kooyman 1986), as well as detailed studies of single species, i.e., California sea lion (Feldkamp 1985), northern elephant seal (Le Boeuf et al. 1988), and a year-round investigation of the Weddell seal (M.A. Castellini, R.W. Davis and G.L. Kooyman, in prep.) have been done.

13.3.1 Weddell Seal

In the most recent study on the Weddell seal, seasonal and geographic comparisons were made. It was found that there is a shift in dive depths from spring (October to December) to summer (January). In the spring and early summer many of the dive depths are from 350 to 450 m (Fig. 13.7a). By late summer this has shifted to depths between 50 and 200 m (Fig. 13.7b). This shift in dive depths

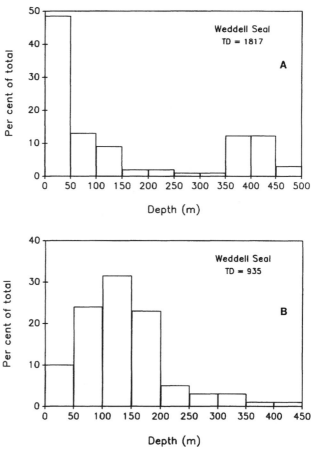

Fig. 13.7. A Frequency of dive depths of a Weddell seal during October through December. *B* Frequency of occurrence of dive depths of a Weddell seal during January. (M.A. Castellini, R.W. Davis and G.L. Kooyman in prep.)

reflects a change in preferred hunting depths that is most apparent from the patterning of dives from individual seals (Fig. 13.8a, b).

In both spring and summer, dives are grouped into bouts with several hours' rest on sea ice between forays. The difference in dive depths seems to have little affect on dive durations of these presumed feeding dives, which range from 5 to 25 min (Fig. 12.1). There is little or no evidence of a diurnal pattern of dive depth and duration. This is distinct from the northern elephant seal.

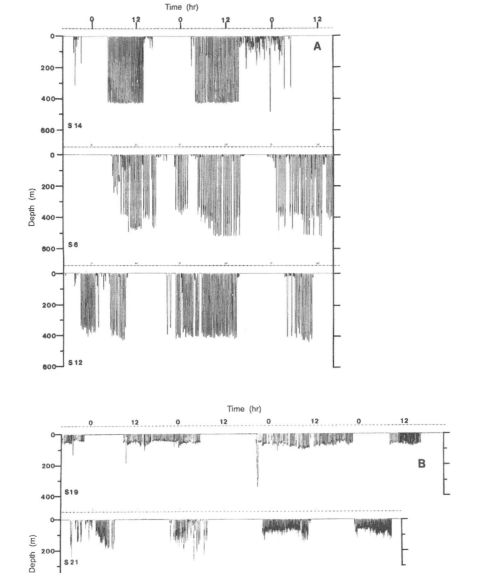

Fig. 13.8. A Three-day dive records of three Weddell seals during November and December. *B* Four-day dive records of four Weddell seals during January. (M.A. Castellini, R.W. Davis and G.L. Kooyman in prep.)

13.3.2 Northern Elephant Seal

In the northern elephant seal the dive pattern of females soon after a 2-month fast, in which the seal has nursed and weaned a pup, is one of continuous diving. In these cold temperate latitudes there is a distinct diel light cycle, in contrast to high latitude cycles where the Weddell seal was studied. Matched to this light cycle is an apparent diurnal pattern in both dive depth and duration (Fig. 13.9). The species is also a truly deep submersible in which dives are seldom <200 m and range mainly between 200 to 700 m (Fig. 13.10). Since only about 10% of

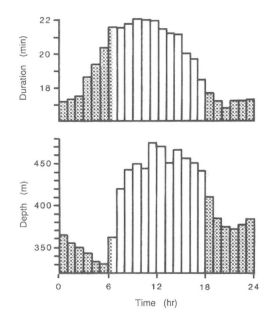

Fig. 13.9. Diurnal variation in average depth and duration for dives of female northern elephant seals during about 10 days of continuous diving. (After Le Boeuf et al. 1988)

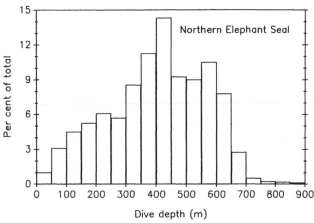

Fig. 13.10. Frequency of occurrence of dive depths in female northern elephant seals. Total of dives was 10,036. (After Le Boeuf et al. 1988)

the time is spent at the surface there is nearly constant exposure to 20 or more atmospheres of pressure. This was noted and discussed in Chapter 4.

These high levels of pressure are made up of consistent dive durations ranging between 11 and 26 min with a skewed distribution towards dives ranging up to about 40 min (Fig. 13.2). It is interesting, as a final remark, that the predicted ADL for these females is about 20 min (Table 12.1). This means that about 37% of the dives exceed the predicted ADL. In comparison, the predicted ADL of Weddell seals is exceeded <5% of the time (Fig. 12.1). The poor correlation in the northern elephant seal may be in part due to a wider range of diving behavior. For example, almost all dives by Weddell seals appeared to be for hunting, in which swim rates and MR are at a consistent level, whereas it has been suggested that during some dives northern elephant seals sleep (Le Boeuf et al. 1986). This seems logical on the basis that all mammals sleep, and since the seals dive continuously it seems likely that during some dives they are sleeping. Further substantiation of this hypothesis is that during several days of velocity measurements in one seal, it was found that during parts of some dives the swim rate was zero (P.J. Ponganis in prep.). This indicates that the seal was drifting at depth. If the animals sleep, then the MR would be much lower for some dives, which would extend the ADL, perhaps to 30 min (Le Boeuf et al. 1988).

13.3.3 Fur Seals

Distinct from the steady, continuous and deep diving of the northern elephant seal, or the bout diving at rather random times of the Weddell seal, female fur seals are generally shallow divers, as exemplified by the antarctic fur seal (Fig. 13.11), that adhere to a strictly nocturnal schedule (Fig. 13.12). There are exceptions such as the northern fur seal in which the frequency distribution of dive depths is bimodal (Fig. 13.13). This is due to a few females that are deep diving (150–200 m) specialists. Interestingly, these specialists dive during the day as well and show little variation in their day and night dive depths (Gentry et al. 1986). The dive durations of fur seals are relatively short. They are usually <2 min and they seldom exceed 4 min.

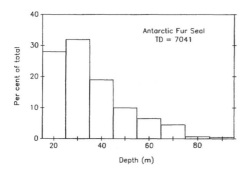

Fig. 13.11. Frequency of occurrence of dive depths in the antarctic fur seal, *Arctocephalus gazella*. (After Kooyman et al. 1986 a)

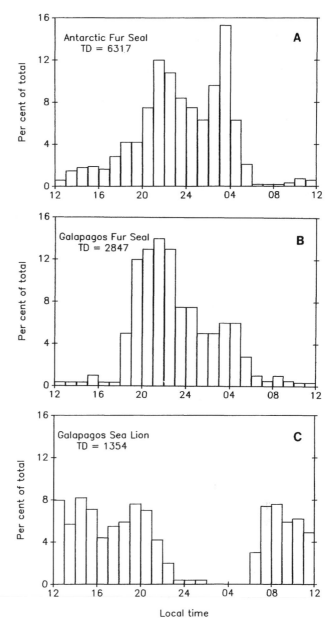

Fig. 13.12. Diurnal variation in proportion of dives that occur per hour of the day for two species of female fur seals and one species of female sea lion. (After Gentry et al. 1986 a)

163

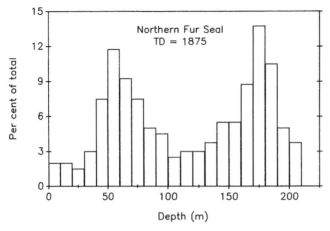

Fig. 13.13. Frequency of occurrence of dive depths for seven northern fur seal females. (After Gentry et al. 1986 b)

13.3.4 Sea Lions

The closely related, but larger sea lions are similar in dive durations and depths. Most dive durations last about 1 to 2 min, seldom exceeding 6 min (Fig. 13.14). Thus, most dives are less than their predicted ADL (Table 12.1). The depths are usually about 40 m (Fig. 13.15), rarely exceeding 100 m; and the deepest dive

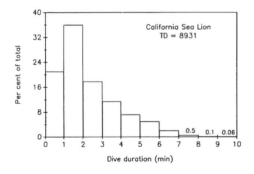

Fig. 13.14. Frequency of occurrence of dive durations in the California sea lion. (After Feldkamp 1985)

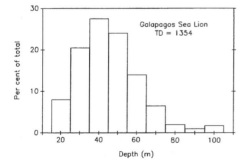

Fig. 13.15. Frequency of occurrence of dive depths in the Galapagos sea lion, Z. c. wol-lebaeki. (After Kooyman and Trillmich 1986)

measured was only 275 m (Table 4.1). Different from fur seals, this group is diurnal (Fig. 13.12), or dives both day and night (Feldkamp 1985). Similarly to the fur seals, the sea lions swim almost constantly for the duration of their time away from the colony. Thus, at least during the nursing cycle, these females may not be continuous divers, but they are continuous swimmers. These various attributes of the California sea lion are especially important to note because, similarly to the Weddell seal, it has been a subject of common choice for physiological studies.

13.3.5 Cetaceans

Although whales have the most extreme adaptations to the marine environment of any tetrapod, they are the least known physiologically and behaviorally. This is not due to lack of desire, but because their natural history does not lend itself to techniques so far developed for behavioral studies. For example, since none has a habit of occasionally coming ashore, as elephant seals and sea turtles do, they must be captured and released at sea. As a result, only one pilot study of diving has, to my knowledge, been completed. In two common dolphins, *Delphinus delphis*, which were tracked for 28 and 6.5 h, dive depths were measured (Evans 1971). It was found that the dolphins dive little in day time. Mean depth in this study was 64 m with a maximum of 260 m. There was no mention of durations.

13.4 Conclusions and Summary

The marked lack of information about whales is frustrating because of all aquatic animals they are the most modified. They may represent the ultimate in extreme adaptations to a way of aquatic life among higher vertebrates. Our curiosity motivates us to try to explain the function of structures in this group for which we have no or little data. One example that relates closely to dive behavior has been the attempt to explain a robust structure unique to some groups of whales. The spermaceti organ makes up about 30% of the body length of the sperm whale, the species in which it is the most developed. It may weigh 12 tons and contain 2.5 tons of oil (Clarke 1979). Combined with a detailed description of its structure, Clarke (1970, 1979) attempted to explain the function as an important bouyancy organ. The proposal was tantamount to suggesting that the sperm whale was a hot-blooded angler fish which waits at depth for prey to approach. The organ assisted by establishing neutral buoyancy and saving energy. This explanation is implausible according to Ridgway (1971) and Norris and Harvey (1972). More likely is the suggestion that it may function as a sound-producing and possibly -amplifying organ (Norris 1972). Norris and Mohl (1983) even go so far as to propose that it may have evolved to such a degree that it is a powerful sonic "stun gun". Recent evidence, nevertheless, does not support this hypothesis, since sonic tests within the frequency range and amplitude produced by whales had little effect on fish (Zagaeski 1987 b). Such ideas that emanate from imaginative thinking about new organs make the prospect of studying whales exciting, to say the least. Other novel organs of cetacea are mentioned also in Chapter 14.

From the more solid data base of other taxa of divers I have shown in this chapter that dive patterns of all species show a high degree of routine behavior. Some fit within the physiological predictions of the effects of compression and the ADL. This is true for the phocids and otariids, as well as perhaps for small penguins. Several groups, however, do not agree with present models either of adaptations to pressure, O_2 management, or aerobic dive limits. The dive durations of yellow-bellied sea snakes are far too long to fit any present model of ADL. Because of the large discrepancy they may be one of the best models for investigating hypometabolism during diving. Related to this proposition more needs to be learned about whether the snakes feed at depth or near the surface, the nature of feeding tactics, and the level of energy expenditure while hunting.

Another reptile, the leatherback sea turtle, fails to conform to its calculated ADL. The dive durations are much shorter than prediction. Again, more needs to be learned about the energetics of foraging. Reptiles seem to be a problem in the prediction of ADL, which in general is based on physiological properties of mammals.

Other problem species in regard to the predictive value of the ADL are the emperor penguin and northern elephant seal, both of which too frequently exceed their predicted ADL to make it an acceptable limit if one subscribes to the hypothesis that most dives are aerobic. The northern elephant seal's behavior might be explained by virtue of its sleeping during some of the longer dives. At present, there is no explanation for that of the penguins.

Another fundamental problem in understanding the energetics of foraging dives is the reason for many routine short dives rather than a few long submersions with long recoveries. An attempt to explain this on the basis of recovery physiology alone showed that for a Weddell seal to return to normal blood acid base levels would require considerable surface time (Fig. 13.16). Using this criterion it was calculated that if a seal made six dives of 15 min duration, each requiring a 4-min recovery, a typical pattern borne out by Fig. 13.8a, b, then 79% of the time would be spent submerged and hunting (Kooyman et al. 1980). In comparison, 70 min of recovery was allowed for a 45-min dive (Fig. 13.16) to yield 39% of the time diving. However, no consideration was given to travel time in these estimates.

If the seals were diving to obtain prey found at 300 m, then based on descent and ascent rates of 50 m min^{-1} (Kooyman 1968) there would be 12 min of travel in a 15-min dive. The bottom time would then be only 20% of the dive time and 16% of the total surface and dive time. Perhaps I have used a worst possible case scenario in that they may descend more directly and rapidly for the deeper dives. It is not likely that descent and ascent rates would be more than about 100 m min^{-1} (see Chap. 11 – Hydrodynamics), which would result in a bottom time of about 47% compared to about 34% for the 45-min dive, if travel time is accounted for in the latter as well. This does seem to give an apparent advantage to

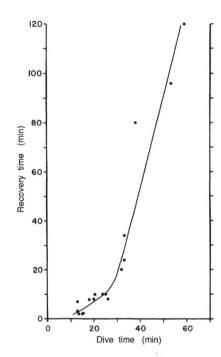

Fig. 13.16. Recovery time required for dives of various durations in Weddell seals. Recovery was considered complete when arterial lactic acid had returned to pre-dive levels. (After Kooyman et al. 1980)

serial, short dives, but not as great as originally proposed. No doubt many other factors such as the greater drain on internal resources incurred by extended dives are involved. One of these resources may be carbohydrates. This fuel is not readily available from the diet, nor stored in large quantities in the body. If it is used abundantly during extended dives, it may be depleted over the course of several serial dives of this type. In conclusion, the behavioral observations add a broader perspective to our physiological models and in many instances force us towards uncertainties about our physiological perceptions and toward major revisions of our hypotheses.

Chapter 14

Final Remarks

A consistent premise throughout this book has been the separation or differentiation between forced submersions and dives. This has been justified for psychogenic reasons, experimental protocol, and results. Nevertheless, in a recent review, Blix (1987) has proposed that the difference is a matter of degree more than of substance. This would mean that extended dives are essentially the same as forced submersions. In both cases the diver is prepared for a long and, at least in the one case, unknown breath-hold duration. I have doubts that the two are completely comparable. In Table 14.1 I have summarized some of the similarities and qualitative differences among forced submersions, extended dives, and bout dives. In these comparisons major differences are the character of muscle activity, LA accumulation, mental attitude, and possible concentrations of catecholamines. The point of view that there is no qualitative difference and that forced submersion and diving represent a difference of degree only is an interesting and important issue. It pleads for further investigation.

Most of the analyses and experiments on voluntary diving have been done on seals. To understand the broadness of these adaptations and which ones, for example, are specialized functions of endotherms versus ectotherms, and deep divers versus shallow divers, work needs to be done on the other groups. This is especially true of birds, in which much of the laboratory work has used a nondiving, domesticated animal. The ultimate irony is that almost no work has been done on whales, the most modified and completely aquatic of all endotherms.

There have been some gallant efforts to remedy this gap by those working with trained whales in the open ocean (Kanwisher and Ridgway 1983). These studies have just scratched the surface, but point the way for further work to un-

Table 14.1. Similarities and differences among some types of breath-holds. + and − are qualitative indicators of the degree of the variable during breath-hold

	Forced submersion	Dive (bout)	Extended dive
Exercise	+[a]	+ +	+
LA Accumulated	+ +	−	+
Psychogenic	+ + +	+ / −	+
Catecholamines	+ +	− ?	?
Variable O_2 Store	−	+	+ +
Bradycardia	+ + +	+	+ +
Restricted Blood Flow	+ + +	+	+ +

[a] Mostly isometric if effectively restrained.

derstand some of the structures unique to whales. What is the function of the spermaceti organ of the Physeteridae and Ziphiidae? What is the reason for the vestigial character of the internal carotid and the usurpation of its role by the spinal meningeal artery, which derives from a complex thoracic rete mirabile (Nagel et al. 1968)? And what function do the series of sphincter muscles in the terminal bronchioles of all the Delphinidae have (Belanger 1940)?

Clearly there is much that is exciting and important to learn about diving in aquatic vertebrates. The new technologies that are becoming available herald a new era in ethology and physiology of diving. Perhaps we will even be able to redress our lack of understanding of how whales work.

Appendix I

Units of Pressure

Units of pressure in the physiological literature are a potpourri. Among those more or less commonly used are millimeters of mercury, centimeters of water, standard atmosphere and torr. Although the pascal (1 Newton per square meter) has been adopted as the basic unit of pressure in the International System of Units (SI System), physiologists have been reluctant to conform with this convention. For those readers not familiar with all of the units I have constructed the following brief list of equivalents:

$1 \text{ Pa} = 1 \text{ N (Newton) m}^{-2} = 0.0075 \text{ mmHg}$.
$1 \text{ mmHg at } 0 \text{ °C} = 1 \text{ torr} = 133.3 \text{ Pa}$.
$1 \text{ cm H}_2\text{O} = 98 \text{ Pa}$.
$1 \text{ atmosphere} = 1.013 \times 10^5 \text{ Nm}^{-2} = 101.3 \text{ kPa} = 14.7 \text{ psi}$
$= 760 \text{ mmHg} \cong 10 \text{ m sea water depth}$.

Appendix II

Scientific Names

Scientific Name	Common Name
Fish	
Salmo	salmon
Reptiles	
Turtles	
Caretta caretta	loggerhead sea turtle
Chelonia mydas	green sea turtle
Chrysemys picta	painted turtle
Dermochelys coriacea	leatherback sea turtle
Lepidochelys kempi	Kemp's sea turtle
Lepidochelys olivacea	olive Ridley sea turtle
Pseudemys concinna	painted turtle, river cooter
Pseudemys scripta	freshwater turtle, pond slider
Sternothaerus minor	musk turtle
Alligators and Crocodiles	
Alligator mississippiensis	American alligator
Caiman crocodilus	caiman crocodile
Lizards	
Amblyrhynchus cristatus	marine iguana
Iguana iguana	common iguana
Turdus migratorius	Australian monitor lizard
Varanus niloticus	Nile monitor lizard
Snakes	
Acalyptophis peronii	sea snake
Achrochordus javanicus	elephant trunk snake, Java wart snake
Aipysurus laevis	olive sea snake
Coluber constrictor	racer (16 sb supp.)
Hydrophis belcheri	blue banded sea snake
Laticauda colubrina	yellow lipped sea krait
Laticauda laticauda	black banded sea krait
Natrix fasciata	banded water snake
Pelamis platurus	yellow-bellied sea snake
Birds	
Penguins	
Aptenodytes forsteri	emperor penguin
Aptenodytes patagonicus	king penguin

Appendix II (continued)

Scientific Name	Common Name
Eupdyptula minor	little blue penguin
Pygoscelis adeliae	Adelie penguin
Pygoscelis antarctica	chinstrap penguin
Pygoscelis papua	gentoo penguin
Spheniscus demersus	jackass penguin
Spheniscus humboldti	Humboldt penguin
Petrels and Alcids	
Fratercula arctica	Atlantic puffin
Pelecanoides georgicus	South Georgia diving petrel
Uria aalge	common murre
Loons, Cormorants, and Ducks	
Aythya americana	redhead duck
Aythya ferina	pochard duck
Aythya fuligula	tufted duck
Gavia immer	common loon
Gavia stellata	red-throated loon
Phalacrocorax auritus	double-crested cormorant
Passerines and Pigeons	
Cinclus mexicana	dipper, water ouzel
Columba livia	domestic pigeon
Mammals	
Seals	
Cystophora cristata	hooded seal
Halichoerus grypus	gray seal
Leptonychotes weddellii	Weddell seal
Mirounga angustirostris	northern elephant seal
Mirounga leonina	southern elephant seal
Phoca fasciata	ribbon seal
Phoca vitulina	harbor seal
Fur Seals and Sea Lions	
Arctocephalus australis	So. American fur seal
Arctocephalus gazella	Antarctic fur seal
Arctocephalus pusillus	So. African fur seal
Callorhinus ursinus	northern fur seal
Phocarctos hookeri	Hooker's sea lion
Zalophus californianus	California sea lion
Z. c. wollebaeki	Galapagos sea lion
Whales	
Balaenoptera physalus	fin whale
Delphinapterus leucas	beluga
Delphinus delphis	common dolphin
Globicephala melaena	pilot whale

174

Appendix II (continued)

Scientific Name	Common Name
Kogia breviceps	pygmy sperm whale
Lagenorhynchus obliquidens	Pacific white-sided dolphin
Orcinus orca	killer whale
Phocoena phocoena	harbor porpoise
Phocoenoides dalli	Dall's porpoise
Physeter catodon	sperm whale
Pseudorca crassidens	false killer whale
Sotalia guianensis [1]	tucuxi, river dolphin, Guiana dolphin
Stenella attenuata	spotted dolphin
Tursiops truncatus	bottlenose dolphin
Walrus	
Odobenus rosmarus	walrus
Dugongs	
Dugong dugong	dugong
Marine Otter	
Enhydra lutris	sea otter
Lutra felina	gato marino
Terrestrial	
Homo sapiens	man
Canis familiaris	domestic dog

[1] Probably *S. fluviatilis*.

Appendix III

Symbols

Δa–A	arterial/alveolar gas tension difference	EXD	extradural intravertebral vein
Δa–$\bar{v}$	arterial-mixed venous content difference	FFA	free fatty acids
ATA	atmospheres absolute	FOG	fast twitch oxidative glycolytic muscle fibers
ATP	adenosine triphosphate	FR	fineness ratio
AVA	arterial venous anastomoses	FT	fast twitch glycolytic muscle fibers
AVC	anterior vena cava	F_{O_2}	fraction of oxygen
BMR	basal metabolic rate	GFR	glomerular filtration rate
B_{max}	maximum body depth	Hb	hemoglobin
BP	breath-hold breaking point	Hct	hematocrit
BPM	beats per min	HPNS	high pressure nervous syndrome
Ca_{N_2}	arterial nitrogen content	HS	hepatic sinus
CBF	cerebral blood flow	ICG	indocyanine green
C_D	drag coefficient	LA	lactic acid
C_{D_A}	drag coefficient-wetted surface area	LCV	lumbar communicating veins
C_{D_F}	drag coefficient-frontal area	LDH	lactate dehydrogenase
$C_{D_{lam}}$	drag coefficient-laminar flow	mb	myoglobin
$C_{D_{turb}}$	drag coefficient-turbulent flow	mM	millimoles per liter
		MR	metabolic rate
C_{D_v}	drag coefficient-volume to 0.66	$\dot{M}_{O_2}$	oxygen exchange rate per time
C_f	frictional drag coefficient	Pa_{N_2}	partial pressure of N_2 in arterial blood
C_p	pressure drag coefficient		
CNS	central nervous system	Pa_{O_2}	partial pressure of O_2 in arterial blood
CO	cardiac output		
CP	creatine phosphate	PA_{CO_2}	partial pressure of alveolar CO_2
CT_{bg}	cost of transport in burst and glide swimming		
CT_{cs}	cost of transport for continuous stroke swimming	P_{CO_2}	partial pressure of CO_2
		P_{O_2}	partial pressure of O_2
D_f	frictional drag	P_{N_2}	partial pressure of nitrogen
D_p	profile drag	$P\bar{v}_{O_2}$	mixed venous partial pressure of O_2
D_T	total drag	PCS	post-caval sphincter
EEG	electroencephalogram	PVC	posterior vena cava

177

PVR	peripheral vascular resistance	SV	stroke volume
		SWB	shallow water blackout
P_{50}	blood oxygen half saturation pressure	TLC	total lung capacity
		u	viscosity
R	range	v	kinematic viscosity
R_{max}	maximum range	V	velocity
Re	Reynolds number	V_{max} or Ve	maximum sustainable velocity
RBC	red blood cells		
RBF	renal blood flow	V_c	average velocity
RMR	resting metabolic rate	V_o	optimal cruising speed
RQ	respiratory quotient	VC	vital capacity
RV	residual capacity	$\dot{V}/\dot{Q}$	ventilation/perfusion ratio
SDA	specific dynamic action	$\dot{V}_{O_2}$	O_2 consumption per time
SMR	standard metabolic rate		
ST	slow twitch oxidative muscle fibers		

References

Ackerman RA, White FN (1979) Cyclic carbon dioxide exchange in the turtle, *Pseudemys scripta*. Physiol Zool 52:378–389

Adams NJ (1987) Foraging range and chick feeding rates of King Penguins, *Aptenodytes patagonia*, at Marion Island during summer. J Zool 212:475–482

Albers C (1974) Carbon dioxide and the utilization of oxygen. In: Nahas G, Schaefer KE (eds) Carbon dioxide and metabolic regulations. Springer, Berlin Heidelberg New York Tokyo, pp 144–149

Aleksiuk M, Frohlinger A (1971) Seasonal metabolic organization in the muskrat (*Ondatra zibethica*). I. Changes in growth, thyroid activity, brown adipose tissue, and organ weights in nature. Can J Zool 49:1143–1154

Aleyev YUG (1977) Nekton. The Hague, Dr. W. Junk, pp 417

Andersen HT (1959) Depression of metabolism in the duck during diving. Acta Physiol Scand 46:234–239

Andersen HT (1961) Physiological adjustments to prolonged diving in the American alligator (*Alligator mississipiensis*). Acta Physiol Scand 53:23–45

Andersen HT (1966) Physiological adaptations in diving vertebrates. Physiol Rev 46:212–243

Angell-James JE, Elsner R, Daly MD (1981) Lung inflation: Effects on heart rate, respiration, and vagal afferent activity in seals. Am J Physiol 240:H190–H198

Aschoff S, Pohl H (1970) Rhythmic variation in energy metabolism. Fed Proc 29:1541–1552

Astrand P, Rodahl K (1977) Textbook of work physiology. McGraw–Hill, New York, 681 pp

Au D, Weihs D (1980) At high speeds dolphins save energy by leaping. Nature (Land) 284:548–550

Baldwin J, Jardel J-P, Montague T, Tomkin R (1984) Energy metabolism in penguin swimming muscles. Mol Physiol 6: 33–42

Barcroft J, Poole LT (1927) The blood in the spleen pulp. J Physiol 64:23–29

Barcroft J, Stephens JG (1927) Observations upon the size of the spleen. J Physiol 64:1–22

Bartholomew GA (1977) Energy metabolism. In: Gordon MS (ed) Animal physiology: Principles and adaptations. Macmillan, New York, pp 57–110

Baudinette RV, Gill P (1985) The energetics of ‚flying‘ and ‚paddling‘ in water: Locomotion in penguins and chicks. J Comp Physiol 155:373–380

Behnke AR (1945) Decompression sickness incident to deep-sea diving and high altitude ascent. Medicine 24:381–402

Behnke AR, Thomson RM, Motley EP (1935) The psychologic effects from breathing air at 4 atmospheres pressure. Am J Physiol 112:554–558

Belanger LF (1940) A study of the histological structure of the respiratory portion of the lungs of aquatic mammals. Am J Anat 67:437–461

Belkin DA (1963) Anoxia: Tolerance in reptiles. Science 139: 492–493

Belkin DA (1964) Variations in heart rate during voluntary diving in the turtle, *Pseudemys concinna*. Copeia 1964:321–330

Belkin DA (1968) Anaerobic brain function: Effects of stagnant and anoxic anoxia on persistence of breathing in reptiles. Science 162:1017–1018

Bello MA, Roy RR, Martin TP, Goforth HW, Edgerton VR (1985) Axial musculature in the dolphin (*Tursiops truncatus*): Some architectural and histochemical characteristics. Mar Mamm Sci 1:324–336

Bennett AF (1973) Blood physiology and oxygen transport during activity in two lizards, *Varanus gouldii* and *Sauromalus hispidus*. Comp Biochem Physiol 46:673–690

Bennett AF, Dawson WR (1976) Metabolism. In: Gans C, Dawson WR (eds) Biology of the reptilia. Academic Press, London, pp 127–213

Bennett PB (1975) Inert gas narcosis. In: Bennet PB, Elliott DH (eds) The physiology and medicine of diving and compressed air work. Williams and Wilkins, Baltimore pp 207–229

Berkson H (1966) Physiological adjustments to prolonged diving in the Pacific green turtle *(Chelonia mydas agassizii)*. Comp Biochem Physiol 18:101–119

Berkson H (1967) Physiological adjustments to deep diving in the Pacific green turtle *(Chelonia mydas)*. Comp Biochem Physiol 21:507–524

Bert P (1869) Sur les differences dans la resistance a l'asphyxie que presentent divers animaux Societe de Biologie, Paris comptes rendus hebonda daires des seances et memoires, pp 186–189

Blake RW (1983 a) Energetics of leaping in dolphins and other aquatic animals. J Mar Biol Assoc 63:61–70

Blake RW (1983 b) Fish locomotion. Cambridge University Press, Cambridge, 208 pp

Blake RW (1983 c) Functional design and burst-and-coast swimming in fishes. Can J Zool 61:2491–2494

Blessing MH (1972) Studies on the concentration of myoglobin in the sea–cow and porpoise. Comp Biochem Physiol 41A:475–480

Blessing MH, Hartschen-Niemeyer E (1969) Über den Myoglobingehalt der Herz und Skelettmuskulatur insbesondere einiger mariner Säuger. Z Biol 116:302–313

Blix AS (1985) Diving response of mammals and birds. In: Rey L (ed) Arctic underwater operations: Medical and operational aspects of diving activities in arctic conditions. Graham and Trotman, London, pp 71–79

Blix AS (1987) Diving responses: Fact or fiction. NIPS 2:64–66

Blix AS, Folkow B (1983) Cardiovascular adjustments to diving in mammals and birds. In: Shepherd JT, Abboud FM (eds) Handbook of physiology, Section 2: The cardiovascular system. Am Physiol Soc, Bethesda, pp 917–945

Blix AS, Kjekshus JK, Enge I, Bergan A (1976) Myocardial blood flow in the diving seal. Acta Physiol Scand 96:277–280

Blix AS, Elsner RW, Kjekshus JK (1983) Cardiac output and its distribution through capillaries and A-V shunts in diving seals. Acta Physiol Scand 118:109–116

Bohr C (1897) Bidrag til Svommefuglernes. Pysiologi K Danshe Vidensk Selsk 2

Bond CF, Gilbert PW (1958) Comparative study of blood volume in representative aquatic and non-aquatic birds. Am J Physiol 194:519–521

Bowers CA, Henderson RS (1972) Project deep ops: Deep object recovery with pilot and killer whales. Nav Undersea Center Tech Pap 306:1–86

Bowers WS (1968) Rete mirabile of dolphin: Its pressure-damping effect on cerebral circulation. Science 161:898–899

Bradley SE, Bing RS (1942) Renal function in the Harbor seal *(Phoca vitulina L)* during asphyxial ischemia and pyrogenic hyperemia. J Cell Comp Biochem 19:229–237

Bradley SE, Mudge GH, Blake WD (1954) The renal excretion of sodium, potassium, and water by the harbor seal *(Phoca vitulina L)*: Effect of apnea; sodium, potassium, and water loading; pitressin; and mercurial diuresis. J Cell Comp Physiol 43:1–22

Brauer RW (1975) The high pressure nervous syndrome: Animals. In: Bennett PB, Elliott DH (eds) The physiology and medicine of diving and compressed air work. Williams and Wilkins, Baltimore, pp 231–247

Brody S (1945) Bioenergetics: Special reference to the efficiency complex in domestic animals. Reinhold, New York, 1023 pp

Bron KM, Murdaugh HV Jr, Millen JE, Lenthall R, Raskin P, Robin ED (1966) Arterial constrictor in a diving mammal. Science 152:540–543

Brooks GA (1985) Glycolytic end product and oxidative substrate during sustained exercise in mammals: The "lactate shuttle". In: Gilles R (ed) Circulation, respiration, and metabolism. Springer, Berlin Heidelberg New York Tokyo, pp 208–218

Bryden MM (1972) Body size and composition of elephant seals *(Mirounga leonina)*: Absolute measurements and estimates from bone dimensions. J Zool (Lond) 167:265–276

Bryden MM, Erickson AW (1972) Body size and composition of Crabeater seals *(Lobodon carcinophagus)*, with observations on tissue and organ size in Ross seals *(Ommatophoca rossi)*. J Zool (Lond) 179:235–247

Bryden MM, Lim GHK (1969) Blood parameters of the southern elephant seal *(Mirounga leonina)* in relation to diving. Comp Biochem Physiol 28:139–148

Bunn HF, Forget BG, Ranney HM (1977) Human hemoglobins. Saunders, Philadelphia, 432 pp

180

Burggren WW (1985) Hemodynamics and regulation of central cardiovascular shunts in reptiles. In: Johansen K, Burggren WW (eds) Alfred Benzon Symposium 21. Cardiovascular shunts: phylogentic, ontogenetic and clinical aspects. Munksgaard International Publishers, Ltd. Copenhagen, pp 121–136

Burggren WW, Shelton S (1979) Gas exchange and transport during intermittent breathing in chelonian reptiles. J Exp Biol 82:75–92

Butler PJ (1979) The use of radio telemetry in the studies of diving and flying birds. In: Amlaner CJ, MacDonald DW (eds) A handbook on biotelemetry and radio tracking. Pergamon Press, Oxford, pp 569–577

Butler PJ (1982 a) Respiratory and cardiovascular control during diving in birds and mammals. J Exp Biol 79:195–221

Butler PJ (1982 b) Respiration during flight and diving in birds. In: Addink ADF, Spronk N (eds) Exogenous and endogenous influences on metabolic and neural control. Pergamon Press, Oxford, pp 103–114

Butler PJ, Jones DR (1982) The comparative physiology of diving in vertebrates. In: Lowenstein O (ed) Advances in comparative physiology and biochemistry. Academic Press, New York, pp 179–364

Butler PJ, Woakes AJ (1979) Changes in heart rate and respiratory frequency during natural behavior of ducks with particular reference to diving. J Exp Biol 283–300

Butler PJ, Woakes AJ (1984) Heart rate and aerobic metabolism in Humboldt penguins, *Spheniscus humboldti*, during voluntary dives. J Exp Biol 108:419–428

Butler PJ, Milsom WK Woakes AJ (1984) Respiratory cardiovascular and metabolic adjustments during steady state swimming in the green turtle, *Chelonia mydas*. J Comp Physiol 154:167–174

Castellini MA, Somero GN (1981) Buffering capacity of vertebrate muscle: Correlations with potentials for anaerobic function. J Comp Physiol 143:191–198

Castellini MA, Murphy BJ, Fedak M, Ronald K, Gofton N, Hochachka PW (1985) Potentially conflicting metabolic demands of apnea and exercise in seals. J Appl Physiol 58:392–399

Castellini MA, Costa DP, Huntley A (1986) Hematocrit variation during sleep apnea. Am J Physiol 251:R429–R431

Castellini MA, Davis RW, Kooyman GL (1988) Blood chemistry regulation during repetitive diving in Weddell seals. Physiol. Zool. 61:379–386

Clark BD, Bemis W (1979) Kinematics of swimming of penguins at the Detroit Zoo. J Zool (Lond) 188:411–428

Clarke MR (1970) Function of the spermaceti organ of the sperm whale. Nature (Lond) 228:873–874

Clarke MR (1979) The head of the sperm whale. Sci Amer 240:128–141

Conroy JWH, Twelves EL (1972) Diving depths of the Gentoo Penguins and Blue-Eyed Shag from the South Orkney Islands. Brit Antarct Bull 30:106–108

Costa DP, Kooyman GL (1984) Contribution of specific dynamic action to heat balance and thermoregulation in the sea otter, *Enhydra lutris*. Physiol Zool 57:199–202

Costa DP, Gentry RL (1986) Free-ranging energetics of northern fur seals. In: Gentry RL, Kooyman GL (eds) Fur seals: Maternal strategies on land and at sea. Princeton University Press, Princeton, pp 79–101

Costello RR, Whittow GC (1975) Oxygen cost of swimming in a trained California sea lion. Comp Biochem Physiol 50:645–647

Cousteau JY (1953) The silent world. London Reprint Soc, London, 266 pp

Craig AB Jr (1961) Causes of loss of consciousness during underwater swimming. J Appl Physiol 16:583–586

Craig AB (1968) Depth limits of breath-hold diving. Respir Physiol 5:14–22

Craig AB, Harley AD (1968) Alveolar gas exchanges during breath-hold dives. J Appl Physiol 24:182–189

Craig AB, Pasche A (1980) Respiratory physiology of freely diving harbor seals (*Phoca vitulina*). Physiol Zool 53:419–432

Cross CE, Packer BS, Altman M, Gee JBL, Murdaugh H Jr (1968) The determination of total body exchangeable O_2 stores. J Clin Invest 47:2402–2410

Cross ER (1965) Taravana: Diving syndrome in the Tuamotu diver. In: Rahn H (ed) Diving and the Ama of japan. NAS-NRC, Washington DC, pp 207–220

Croxall JP, Davis RW, O'Connell MJ (1988) Diving patterns in relation to diet in Gentoo and Macaroni penguins at South Georgia. Condor 90:157–167 de Burgh Daly MD, Elsner R, Angell-James JE (1977) Cardiorespiratory control by carotid chemoreceptors during experimental dives in the seal. Am J Physiol 232:H508–H516

Davis RW (1983) Lactate and glucose metabolism in the resting and diving harbor seal (*Phoca vitulina*). J Comp Physiol 153:275–288

Davis RW, Castellini MA, Kooyman GL, Maue R (1983) GFR and hepatic blood flow during voluntary diving in Weddell seals. Am J Physiol 245:743–748

Davis RW, Williams TM, Kooyman GL (1985) Swimming metabolism of yearling and adult harbor seals (*Phoca vitulina*). Physiol Zool 58:590–596

Dawson WR, Bartholomew GA, Bennett AF (1977) A reappraisal of the aquatic specializations of the Galapagos marine iguana (*Amblyrhynchus cristatus*). Evolution 31:891–897

Dietz RS (1962) The sea's deep scattering layers. Sci Amer 207:1–8

Djojosugito AM, Folkow B, Yonce LR (1969) Neurogenic adjustments of muscle blood flow, cutaneous A-V shunt flow and of venous tone during diving in ducks. Acta Physiol Scand 75:377–386

Dolphin WF (1987) Dive behavior and estimated energy expenditure of foraging humpback whales in southeast Alaska. Can J Zool 65:354–362

Dormer KJ, Denn MJ, Stone HL (1977) Cerebral blood flow in the sea lion (*Zalophus californianus*) during voluntary dives. Comp Biochem Physiol 58:11–18

Drent RH, Stonehouse B (1971) Thermoregulatory responses of the Peruvian Penguin, *Spheniscus humboldti*. Comp Biochem Physiol 40:689–710

Dunlap CE (1955) Notes on the visceral anatomy of the giant leatherback turtle (*Dermochelys coriacea Linnaeus*). Bull Tulane Univ Med Fac 14:55–69

Dykes RW (1974a) Factors related to the dive reflex in harbor seals: Respiration, immersion, bradycardia, and lability of the heart rate. Can J Physiol Pharmacol 52:248–258

Dykes RW (1974b) Factors related to the dive reflex in harbor seals: Sensory contributions from the trigeminal region. Can J Physiol Pharmacol 52:259–265

Eckert SA, Nellis DW, Eckert KL, Kooyman GL (1986) Diving pattern in two leatherback sea turtles (*Dermochelys coriacea*) during internesting intervals at Sandy Point, St. Croix, US Virgin Islands. Herpetologica 42:381–388

Eckert SA, Eckert KL, Ponganis PJ, Kooyman GL (1989) Diving and foraging behavior of leatherback sea turtles (*Dermochelys coriacea*) Canad J Zool (In Press)

Eliassen E (1960) Cardiovascular responses to submersion asphyxia in avian divers. Univ Bergen Mat Naturv Ser 2:1–100

Elliott DH, Hallenbeck JM (1975) The pathophysiology of decompression sickness. In: Bennett PB, Elliott DH (eds) The physiology and medicine of diving and compressed air work. Williams and Wilkins, Baltimore, pp 435–455

Elsner RW (1965) Heart rate response in forced versus trained experimental dives in pinnipeds. Hvalrad Skr 48:24–29

Elsner RW (1966) Diving bradycardia in the unrestrained hippopotamus. Nature (Lond) 212:408

Elsner RW (1969) Cardiovascular adjustments to diving. In: Andersen HT (ed) The biology of marine mammals. Academic Press, New York, pp 117–146

Elsner RW (1970) Diving mammals. Sci J 6:68–74

Elsner RW, Ashwell-Erickson S (1982) Maximum oxygen consumption of exercising harbor seals. Physiologist 25:279

Elsner RW, Gooden B (1983) Diving and asphyxia: A comparative study of animals and man. Physiological Society Monograph 40. Cambridge University Press, Cambridge, 175 pp

Elsner RW, Scholander PF (1965) Circulatory adaptations to diving in animals and man. In: Physiology of breath-hold diving and the Ama of Japan. Publ. 1341, Nat Acad Sci, Nat Res Counc, Washington DC, pp 281–294

Elsner RW, Franklin DL, VanCitters RL (1964) Cardiac output during diving in an unrestrained sea lion. Nature (Lond) 153:809–810

Elsner RW, Franklin DL, VanCitters RL, Kinney DW (1966) Cardiovascular defense against asphyxia. Science 153:941–949

Elsner RW, Hammond DD, Parker HR (1970 a) Circulatory responses to asphyxia in pregnant and fetal animals: A comparative study of Weddell seals and sheep. Yale J Biol Med 42:202–217

Elsner RW, Shurley JT, Hammond DD, Brooks RE (1970 b) Cerebral tolerance to hypoxemia in asphyxiated Weddell seals. Respir Physiol 9:287–297

Elsner RW, Hanafee WN, Hammond DD (1971) Angiography of the inferior vena cava of the harbor seal during simulated diving. Am J Physiol 220:1155–1157

Elsner RW, Hammel HT, Heller HC (1975) Combined thermal and diving stresses in the harbor seal, *Phoca vitulina*: A preliminary report. Rapp P-V Reun Cons Int Explor Mer 169: 437–440

Elsner RW, Blix AS, Kjekshus J (1978) Tissue perfusion and ischemia in diving seals. Physiologist 21:33

Evans WE (1971) Orientation behavior of delphinids: Radio telemetric studies. Ann NY Acad Sci 188:142–160

Falke KJ, Hill RD, Qvist J, Schneider RC, Guppy M, Liggins GC, Hochachka PW, Elliott RE, Zapol WM (1985) Seal lungs collapse during free diving: Evidence from arterial nitrogen tensions. Science 229:556–558

Fedak MA (1986) Diving and exercise in seals: A benthic perspective. In: Brubakk A, Kanwisher JW, Sundnes G (eds) Diving in animals and man. Symposium 1985, R Norwegian Soc Sci Lett, Papir, Trondheim

Feldkamp SD (1985) Swimming and diving in the California sea lion, *Zalophus californianus*. Doctoral Dissertation, Scripps Inst Oceanogr, University of California, San Diego, 176 pp

Feldkamp SD (1987) Swimming in the California sea lion: morphometrics, drag and energetics. J Exp Biol 131:117–135

Felger RS, Cliffton K, Regal PJ (1976) Winter dormancy in sea turtles: Independent discovery and exploitation in the Gulf of California by two local cultures. Science 191:283–284

Folkow B, Neil E (1977) Circulation. Oxford University Press, London, 593 pp

Folkow B, Fuxe K, Sonnenschein RR (1966) Responses of skeletal musculature and its vasculature during "diving" in the duck: Peculiarities of the adrenergic vasoconstrictor innervation. Acta Physiol Scand 67:327–342

Folkow B, Nilsson NJ, Yonce LR (1967) Effects of diving on cardiac output in ducks. Acta Physiol Scand 70:347–361

Frair W (1977) Sea turtle red blood cell parameters correlated with carapace length. Physiol Zool 53:394–401

Frair W, Ackman RG, Mrosovsky N (1972) Body temperature of *Dermochelys coriacea*: Warm turtle from cold water. Science 177:791–793

Fujise Y, Hidaka H, Tatsukawa R, Miyazaki N (1985) External measurements and organ weights of five Weddell seals (*Leptonychotes weddelli*) caught near Syowa Station. 85:96–101

Furilla RA, Jones DR (1986) The contribution of nasal receptors to the cardiac response to diving in restrained and unrestrained redhead ducks (*Aythya americana*). J Exp Biol 121:227–238

Gabbott GRJ, Jones DR (1985) Psychogenic influences on the cardiac response of the duck (*Anas platyrhynchos*) to forced submersion. J Physiol 371:71p

Gabrielsen GW (1985) Free and forced diving in ducks: Habituation of the initial dive response. Acta Physiol Scand 123:67–72

Galliano RE, Morgane PJ, McFarland WL, Nagel EL, Catherman RL (1966) The anatomy of the cervicothoracic arterial system in the bottlenose dolphin (*Tursiops truncatus*) with a surgical approach suitable for guided angiography. Anat Rec 155:325–338

Gallivan GJ, Kanwisher JW, Best RC (1986) Heart rates and gas exchange in the Amazonian Manatee (*Trichechus inunguis*). J Comp Physiol 156:415–423

Gatten RE (1981) Anaerobic metabolism in freely diving painted turtles *(Chrysemys picta)*. J Exp Zool 216:377–385

Gaunt GS, Gans C (1969) Diving bradycardia and withdrawal bradycardia in *Caiman crocodilus*. Nature (Lond) 223:207–208

Gentry RL, Costa DP, Croxall JP, David JHM, Davis RW, Kooyman GL, Majluf P, McCann TS, Trillmich F (1986 a) Synthesis and conclusions. In: Gentry RL, Kooyman GL (eds) Fur seals: Maternal strategies on land and at sea. Princeton University Press, New Jersey, pp 220–264

Gentry RL, Kooyman GL (eds) (1986) Fur seals: maternal strategies on land and at se. Princeton Univ Press, New Jersey 291 p

Gentry RL, Kooyman GL, Goebel ME (1986 b) Feeding and diving behavior of northern fur seals. In: Gentry RL, Kooyman GL (eds) Fur seals: Maternal strategies on land and at sea. Princeton University Press, New Jersey, pp 61–78

Gleeson TT (1980) Lactic acid production during field activity in the Galapagos marine iguana, *Amblyrhynchus cristatus*. Physiol Zool 53:157–162

Gollnick PD, Hermansen L (1973) Biochemical adaptations to exercise: Anaerobic metabolism. In: Wilmore JH (ed) Exercise and sport reviews. Academic Press, New York, pp 1–43

Gordon CN (1980) Leaping dolphins. Nature 287:759

Graham JB (1974) Aquatic respiration in the sea snake. Respir Physiol 21:1–7

Graham JB (1987) Surface and Subsurface swimming of the sea snake *Pelamis platurus*. Physiol Zool 127:27–44

Graham SF (1967) Seal ears. Science 155:489

Gucinski H, Bauer RE (1983) Surface properties of porpoise and killer whale skin in vivo. Am Soc Zool 23:959

Guppy M, Hill RD, Schneider RC, Qvist J, Liggins GC, Zapol WM, Hochachka PW (1986) Microcomputer-assisted metabolic studies of voluntary diving of Weddell seals. Am J Physiol 250:R175–R187

Guyton AC (1981) Textbook of medical physiology. Saunders, Philadelphia, 1074 pp

Halasz NA, Elsner R, Garvie RS (1974) Renal recovery from ischemia: A comparative study of harbor seal and dog kidneys. Am J Physiol 227:1331–1335

Hall JD (1970) Conditioning Pacific white striped dolphin, *Lagenorhynchus obliquidens*, for open ocean release. Nav Undersea Center Techn Pap 200:1–16

Harrison RJ, Tomlinson JDW (1963) Anatomical and physiological adaptations in diving mammals. Viewpoints Biol 2:115–162

Harvey EN, McElroy WD, Whiteley AH, Warren GH, Pease DC (1944) Bubble formation in animals. III. An analysis of gas tension and hydrostatic pressure in cats. J Cell Comp Physiol 24:117–132

Heatwole H, Seymour RS, Webster MED (1979) Heart rates of sea snakes diving in the sea. Comp Biochem Physiol 62:453–456

Heezen BC (1957) Whales entangled in deep-sea cables. Deep-Sea Res 4:105–115

Herbert CV, Jackson DC (1985) Temperature effects on the responses to prolonged submergence in the turtle, *Chrysemys picta bellii*. II. Metabolic rate, blood acid–base and ionic changes, and cardiovascular function in aerated and anoxic water. Physiol Zool 58:670–681

Hertel H (1969) Hydrodynamics of swimming and wave-riding dolphins. In: Andersen HT (ed) The biology of marine mammals. Academic Press, New York, pp 31–64

Hill RD, Schneider RC, Liggins GC, Schuette AH, Elliott RL, Guppy M, Hochachka PW, Qvist J, Falke KJ, Zapol WM (1987) Heart rate and body temperature during free diving of Weddell seals. Am J Physiol 253:R344–R351

Hochachka PW (1980) Living without oxygen. Harvard University Press, Cambridge, 181 pp

Hochachka PW (1981) Brain, lung and heart function during diving and recovery. Science 212:509–514

Hochachka PW, Guppy M (1987) Diving mammals and birds. In: Hochachka PW, Guppy M (eds) Metabolic arrest and the control of biological time. Harvard University Press, Cambridge, pp 36–56

Hochachka PW, Murphy B, Liggins GC, Zapol W, Creasy R, Snider M, Schneider R, Qvist J (1979) Unusual maternal-fetal blood glucose concentrations in Weddell seal. Nature 227:388–389

Hochachka PW, Murphy B (1979) Metabolic status during diving and recovery in marine mammals. In: Shaw RD (ed) International review of physiology, environmental physiology III. Park, Baltimore, pp 253–287

Hochachka PW, Somero GN (1984) Biochemical adaptation. Princeton University Press, Princeton, 537 pp

Hochachka PW, Liggins GC, Qvist J, Schneider R, Snider MY, Wonders TR, Zapol WM (1977) Pulmonary metabolism during diving: Conditioning blood for the brain. Science 198:831–834

Hoerner SF (1965) Fluid-dynamic drag: Practical information on aerodynamic drag and hydrodynamic resistance. Hoerner SF (ed). Published by Author, Midland Park, NJ

Hogan MC, Welch HG (1986) The effect of altered arterial oxygen tensions on muscle metabolism in dog skeletal muscle during fatiguing work. Am J Physiol 251:C216–C222

Holloszy JO, Booth FW (1976) Biochemical adaptations to endurance exercise in muscle. Ann Rev Physiol 38:273–291

Hong SK (1988) Man as a breath-holder. Can J Physiol 66:70–74

Hong SK, Ashwell-Erickson S, Gigliotti P, Elsner RW (1982) Effects of anoxia and low pH on organic ion transport and electrolyte distribution in the harbor seal (*Phoca vitulina*) kidney slices. J Comp Physiol 149:19–24

Horvath SM, Chiodi H, Ridgway SH, Azar S (1968) Respiratory and electrophoretic characteristics of hemoglobin of porpoises and sea lion. Comp Biochem Physiol 24:1027–1033

Hudson DM, Jones DR (1986) The influence of body mass on the endurance to restrained submergence in the Pekin duck. J Exp Biol 120:351–367

Hui CA (1983) Swimming in penguins. Doctoral Dissertation, UCLA, 186 pp

Hui CA (1986) The porpoising of penguins: An energy conserving behavior for respiratory ventilation? Can J Zool 65:209–211

Huxley FM (1913) On the reflex nature of apnea in duck in diving. II. The reflex nature of submersion apnea. Q J Exp Physiol 6:147–157

Irving L (1935) The protection of whales from the danger of caisson disease. Science 81:560–561

Irving L (1964) Comparative anatomy and physiology of gas transport mechanisms. In: Fenn WO, Rahn H (eds) Handbook of physiology. Sect 3. Respiration, vol I. Am Physiol Soc, Washington DC, pp 177–212

Jackson DC (1968) Metabolic depression and oxygen depletion in the diving turtle. Apl Physiol 24:503–509

Jackson DC, Heisler N (1982) Plasma ion balance of submerged anoxic turtles at 3 °C: The role of calcium lactate formation. Respir Physiol 49:159–174

Jackson DC, Schmidt-Nielsen K (1966) Heat production during diving in the fresh water turtle, *Pseudemys scripta*. J Cell Physiol 67:225–232

Johannessen CL, Harder JA (1960) Sustained swimming speeds of dolphins. Science 132:1550–1551

Johansen K (1964) Regional distribution of circulating blood during submersion asphyxia in the duck. Acta Physiol Scand 62:1–9

Johansen K, Lenfant C (1972) Oxygen affinity of hemoglobin and red cell acid-base status. In: Rorth M, Astrup P (eds) A comparative approach to the adaptability of O_2–Hb affinity. Alfred Benzon Symposium, Munksgaard, Copenhagen, pp 750–780

Johansen K, Burggren W, Glass M (1977) Pulmonary stretch receptors regulate heart rate and pulmonary blood flow in the turtle, *P. scripta*. Comp Biochem Physiol 58:185–191

Johlin JM, Moreland FB (1933) Studies of the blood picture of the turtle after complete anoxia. J Biol Chem 103:107–114

Jones DR, Fisher HD, McTaggart S, West NH (1973) Heart rate during breath-holding and diving in the unrestrained harbor seal (*Phoca vitulina richardi*) Can J Zool 51:671–680

Jones DR, Furilla RA (1987) The anatomical, physiological, behavioral, and metabolic consequences of voluntary and forced diving. In: Seller TJ (ed) Bird respiration. Vol II. CRC Press, Inc Boca Raton, Florida, pp 75–125

Jones DR, Holeton GF (1972) Cardiac output of ducks during diving. Comp Biochem Physiol 41:639–645

Jones DR, Milsom WK, Gabbott GRJ (1982) Role of central and peripheral chemoreceptors in diving responses of ducks. Am J Physiol 243:R537–R545

Jones DR, Bryan RM Jr, West NH, Lord RH, Clark B (1979) Regional distribution of blood flow during diving in the duck (*Anas platyrhynchos*). Can J Zool 57:995–1002

Kanwisher JW, Gabrielsen GW (1985) The diving response in man. In: Rey L (ed) Arctic underwater operations: Medical and operational aspects of diving activities in arctic conditions. Graham and Trotman, London, pp 81–95

Kanwisher JW, Ridgway SH (1983) The physiological ecology of whales and porpoises. Sci Am 248:111–119

Kanwisher JW, Gabrielsen G, Kanwisher N (1981) Free and forced diving habits in birds. Science 211:717–719

Kerem D, Elsner R (1973) Cerebral tolerance to asphyxial hypoxia in the dog. Am J Physiol 225:593–600

Kerem D, Kooyman GL, Schroeder JP, Wright JJ, Drabek CM (1972) Hyperbaric oxygen–induced seizure in a marine mammal, the seal. Am J Physiol 222:1322–1325

Kerem DH, Hammond DD, Elsner R (1973) Tissue glycogen levels in the Weddell seals (*Leptonychotes weddelli*): A possible adaptation to asphyxial hypoxia. Comp Biochem Physiol 45:731–736

Kjekshus JK, Blix AS, Elsner R, Hol R, Amundsen E (1982) Myocardial blood flow and metabolism in the diving seal. Am J Physiol 242:R97–R104

Kleiber M (1961) The fire of life. Wiley and Sons, New York, 454 pp

Kolb PM, Norris KS (1982) A harbor seal, *Phoca vitulina richardi*, taken from a sablefish trap. Calif Dept Fish Game 68:123–124

Kooyman GL (1965) Techniques used in measuring diving capacities of Weddell seals. Polar Rec 12:391–394

Kooyman GL (1966) Maximum diving capacities of the Weddell seal (*Leptonychotes weddelli*). Science 151:1553–1554

Kooyman GL (1968) An analysis of some behavioral and physiological characteristics related to diving in the Weddell seal. Antarct Biol Ser 11:227–261

Kooyman GL (1972) Deep diving behavior and effects of pressure on reptiles, birds, and mammals. In: S E B Symposium No. 26. The effects of pressure on organisms. London, p 295–313

Kooyman GL (1973) Respiratory adaptations in marine mammals. Am Zool 13:457–478

Kooyman GL (1974) Behavior and physiology of diving. In: Stonehouse B (ed) The biology of penguins. MacMillan, London, pp 115–137

Kooyman GL (1981) Weddell seal: Consummate diver. Cambridge University Press, Cambridge, 135 pp

Kooyman GL (1982) How marine mammals dive. In: Taylor CR, Johansen K, Bolis L (eds) A companion to animal physiology. Cambridge University Press, Cambridge, pp 151–160

Kooyman GL (1985) Physiology without restraint in diving mammals. Mar Mamm Sci 1:166–178

Kooyman GL (1987 a) Free diving in vertebrates: Common denominators. In: Dejours P (ed) Comparative physiology of environmental adaptations. Vol 2, 8th Conf Eur Soc Comp Physiol Biochem, Strasbourg, 1986. Karger, Basel, Switzerland, pp 27–37

Kooyman GL (1987 b) A reappraisal of diving physiology: Seals and penguins. In: Dejours P, Bolis L, Taylor CR, Weibel ER (eds) Comparative physiology: Life in water and on land, vol 9. Springer, Berlin Heidelberg New York Tokyo, pp 459–469

Kooyman GL, Campbell WB (1973) Heart rate in freely diving Weddell seals (*Leptonychotes weddelli*). Comp Biochem Physiol 43:31–36

Kooyman GL, Davis RW (1987) Diving behavior and performance, with special reference to penguins. In: Seabirds: Feeding biology and role in marine ecosystems. Cambridge University Press, London, pp 63–75

Kooyman GL, Gentry RL (1986) Diving behavior of South African fur seals. In: Gentry RL, Kooyman GL (eds) Fur seals: Maternal strategies on land and at sea. Princeton University Press, Princeton, pp 142–152

Kooyman GL, Sinnett EE (1979) Mechanical properties of the harbor porpoise lung. Respir Physiol 36:287–300

Kooyman GL, Sinnett EE (1982) Pulmonary shunts in harbor seals and sea lions during simulated dives to depth. Physiol Zool 55:105–111

Kooyman GL, Trillmich F (1986) Diving behavior of Galapagos sea lions. In: Gentry RL, Kooyman GL (eds) Fur seals: Maternal strategies on land and at sea. Princeton University Press, New Jersey, pp 209–219

Kooyman GL, Drabek CM, Elsner R, Campbell WB (1971 a) Diving behavior of the Emperor Penguin, *Aptenodytes forsteri*. Auk 88:775–795

Kooyman GL, Kerem DH, Campbell WB, Wright JJ (1971 b) Pulmonary function in freely diving Weddell seals (*Leptonychotes weddelli*). Respir Physiol 12:271–282

Kooyman GL, Schroeder JP, Denison DM, Hammond DD, Wright JJ, Bergman WP (1972) Blood N2 tensions of seals during simulated deep dives. Am J Physiol 223:1016–1020

Kooyman GL, Kerem DH, Campbell WB, Wright JJ (1973 a) Pulmonary gas exchange in freely diving Weddell seals (*Leptonychotes weddelli*). Respir Physiol 17:283–290

Kooyman GL, Schroeder JP, Greene DG, Smith VA (1973 b) Gas exchange in penguins during simulated dives to 30 and 68 m. Am J Physiol 225:1467–1471

Kooyman GL, Gentry RL, Urquhart DL (1976) Northern fur seal diving behavior: A new approach to its study. Science 193: 411–412

Kooyman GL, Wahrenbrock EA, Castellini MA, Davis RA, Sinnett EE (1980) Aerobic and anaerobic metabolism during diving in Weddell seals: Evidence of preferred pathways from blood chemistry and behavior. J Comp Physiol 138:335–346

Kooyman GL, Davis RW, Croxall JP (1986) Diving behavior of Antarctic fur seals. In: Gentry RL, Kooyman GL (eds) Fur seals: maternal strategies on land and at sea. Princeton University Press, Princeton, pp 115–125

Kooyman GL, Davis RW, Croxall JP, Costa DP (1982) Diving depths and energy requirements of King Penguins. Science 217: 726–727

Kooyman GL, Castellini MA, Davis RW, Maue RA (1983) Aerobic dive limits in immature Weddell seals. J Comp Physiol 151:171–174

Koschier FJ, Elsner RW, Holleman DF, Hong SK (1978) Organic anion transport by renal cortical slices of harbor seals (*Phoca vitulina*). Comp Biochem Physiol 60:289–292

Kramer DL (1988) The behavioral ecology of air breathing by aquatic animals. Can J Zool, 66:89–94

Kramer K, Deetjen P (1964) Oxygen consumption and sodium reabsorption in the mammalian kidney. In: Dickens F, Neil E (eds) Oxygen in the animal organism. Pergamon Press, New York, pp 411–431

Kramer K, Luft UC (1951) Mobilization of red cells and oxygen from the spleen in severe hypoxia. J Physiol 165:215–218

Kramer MO (1965) Hydrodynamics of the dolphin. Adv Hydrosci 2: 111–130

Lane RAB, Morris RJH, Sheedy JW (1972) A haemotological study of the southern elephant seal, *Mirounga leonina* (Linn.). J Comp Physiol 42A:841–850

Landis AT (1965) New high pressure research animal? Nav Undersea Center Tech Pap 6:21

Lang TG (1975) Speed, power, and drag measurements of dolphins and porpoises. In: Wu TY-T, Brokaw CJ, Brennen C (eds) Swimming and flying in nature. Plenum Press, New York, pp 553–572

Langlois P, Richet C (1898 a) Des gas expirés par les canards plongés dans l'eau. C R Soc Biol 5:483–486

Langlois P, Richet C (1898b) Dosage des gaz dans l'asphyxie du canard. C R Soc Biol 5:718–719

Lanphier EH, Rahn H (1963 a) Alveolar gas exchange during breath-hold diving. J Appl Physiol 18:471–477

Lanphier EH, Rahn H (1963b) Alveolar gas exchange during breath-holding with air. J Appl Physiol 18:478–482

Lapennas GN, Lutz PL (1982) Oxygen affinity of sea turtle blood. Respir Physiol 48:56–74

Lasiewski R, Dawson WR (1967) A re-examination of the relation between standard metabolic rate and body weight in birds. Condor 69:13–23

Lavigne DM, Barchard W, Innes S, Oritsland NA (1982) Pinniped bioenergetics. In: Clark GS (ed) Mammals in the seas: Small cetaceans, seals, sirenians and others, vol IV, FAO Fisheries Series 5, Rome, pp 191–235

Lavigne DM, Innes S, Worthy GAJ, Kovacs KM, Schmitz OJ, Hickie JP (1986) Metabolic rates of seals and whales. Can J Zool 64: 279–284

Lawrie RA (1950) Some observations on factors affecting myoglobin concentrations in muscle. J Agric Sci 40:56–366

Le Boeuf BJ, Costa DP, Huntley AC, Kooyman GL, Davis RW (1986) Pattern and depth of dives in two northern elephant seals. J Zool (Lond) 208:1–7

Le Boeuf BJ, Costa DP, Huntley AC, Feldkamp SD (1988) Continuous, deep diving in female Northern seals, *Mirounga angustirostris*. Can J Zool 66:446–458

Leith D (1976) Comparative mammalian respiratory mechanics. Physiologist 19:485–510

Lenfant C (1969) Physiological properties of blood of marine mammals. In: Andersen HT (ed) The biology of marine mammals. Academic Press, New York, pp 95–116

Lenfant C, Kooyman GL, Elsner R, Drabek CM (1969 a) Respiratory function of the blood of the Adelie penguin (*Pygoscelis adeliae*). Am J Physiol 216:1598–1600

Lenfant C, Elsner T, Kooyman GL, Drabek CM (1969b) Respiratory function of the blood of the adult and fetus Weddell seal, *Leptonychotes weddelli* (1). Am J Physiol 216:1595–1597

Lenfant C, Johansen K, Torrance JD (1970) Gas transport and oxygen storage capacity in some pinnipeds and the sea otter. Respir Physiol 9:277–286

Liggins GC, Qvist J, Hochachka PW, Murphy BJ, Creasy RK, Schneider RC, Snider MT, Zapol WM (1980) Fetal cardiovascular and metabolic responses to simulated diving in the Weddell seal. J Appl Physiol 49:424–430

Lin YC, Matsura DT, Whittow GC (1972) Respiratory variation of heart rate in the California sea lion. Am J Physiol 222: 260–264

Lishman GS, Croxall JP (1983) Diving depths of the Chinstrap Penguin, *Pygoscelis antarctica*. Brit Antarct Bull 61:21–25

Lutz PL, Bentley TB (1985) Respiratory physiology of diving in the sea turtle. Copeia 1985:671–679

Lutz PL, Longmuir IS, Tuttle JV, Schmidt-Nielsen K (1973) Dissociation curve of bird blood and effect of red cell oxygen consumption. Respir Physiol 17:269–275

Lutz PL, LaManna JC, Adams MR, Rosenthal M (1980) Cerebral resistance to anoxia in the marine turtle. Respir Physiol 41: 241–251

Lutz PL, McMahon P, Rosenthal M, Sick TJ (1984) Relationships between aerobic and anaerobic energy production in turtle brain in situ. Am J Physiol 247:R740–R744

Schmidt-Nielsen B, Murdaugh HV, O'Dell R, Bacsanyi J (1959) Urea excretion and diving in the seal (*Phoca vitulina L.*). J Cell Comp Physiol 53:393–411

Schmidt–Nielsen K (1983) Animal Physiology: Adaptation and Environment. Cambridge University Press, Cambridge, 619 pp

Scholander PF (1940) Experimental investigations on the respiratory function in diving mammals and birds. Hvalradets Skr Nor Vidensk, Akad Oslo 22:1–131

Scholander PF, Irving L, Grinnell SW (1942) Aerobic and anaerobic changes in seal muscle during diving. J Biol Chem 142:431–440

Schorger AW (1947) The deep diving of the Loon and the Old-Squaw and its mechanism. Wilson Bull 59:151–159

Severinghaus JW (1974) Carbon dioxide transport. In: Nahas G, Schaefer KE (eds) Carbon dioxide and metabolic regulations. Springer, Berlin Heidelberg New York Tokyo, pp 138–143

Seymour RS (1978) Gas tensions and blood distribution in sea snakes at surface pressure and at simulated depth. Physiol Zool 51:388–407

Seymour RS (1979) Blood lactate in free-diving snakes. Copeia 1979:494–497

Seymour RS (1982) Physiological adaptations of aquatic life. In: Gans C, Hough FH (eds) Biology of reptilia. Academic Press, New York, pp 1–51

Seymour RS, Webster MED (1975) Gas transport and blood acid-base balance in diving sea snakes. J Exp Zool 191:169–182

Shepherd JT (1983) Circulation to skeletal muscle. In: Sheperd JT, Abboud FM (eds) Handbook of physiology: The cardiovascular system. Am Physiol Soc, Bethesda, pp 319–370

Shepherd JT, Vanhoutte PM (1975) Skeletal muscle blood flow: Neurogenic determinants. In: Zellis R (ed) The peripheral circulations. Grune and Stratton, New York, pp 3–56

Sick TJ, Rosenthal M, LaManna JC, Lutz PL (1982) Brain potassium ion homeostasis, anoxia, and metabolic inhibition in turtles and rats. Am J Physiol 243:R281–R288

Simpson JG, Gilmartin WG, Ridgway SH (1970) Blood volume and other hematologic values in young elephant seals (*Mirounga angustirostris*). Am J Vet Res 31:1449–1452

Sinnett EE, Kooyman GL, Wahrenbrock EA (1978) Pulmonary circulation of the harbor seal. J Appl Physiol 45:718–727

Skinner NS (1975) Skeletal-muscle blood flow: Metabolic determinants. In: Zellis R (ed) The peripheral circulations. Grune and Stratton, New York, pp 57–78

Sleet RB, Sumich JL, Weber LJ (1981) Estimates of total blood volume and total body weight of sperm whale (*Physeter catodon*). Can J Zool 59:567–570

Smith EN, Allison RD, Crowder WE (1974) Bradycardia in free ranging American alligator. Copeia 1974:770–772

Snow DH, Harris RC (1985) Thoroughbreds and greyhounds: Biochemical adaptations in creatures of nature and of man. In: Gilles R (ed) Circulation, respiration, and metabolism: Current comparative approaches. Springer, Berlin Heidelberg New York Tokyo, pp 227–239

Snyder GK (1983) Respiratory adaptations in diving mammals. Respir Physiol 54:269–294

Sparks HV, Belloni FL (1978) The peripheral circulation: Local regulation. Ann Rev Physiol 40:67–92

Spencer MP, Campbell SD (1968) Development of bubbles in venous and arterial blood during hyperbaric decompression. Mason Res Clinic Bull 22:11–17

Stahl W (1965) Organ weights in primates and other animals. Science 150:1039–1042

Stanley WC, Gertz EW, Wisneski JA, Neese RA, Morris DL, Brooks GA (1986) Lactate extraction during net lactate release in legs of humans during exercise. J Appl Physiol 60:1116–1120

Stevens GA (1950) Swimming of dolphins. Sci Prog 38:524–525

Stone HL, Gray K, Stabe R, Chandler JM (1973) Renal blood flow in a diving trained sea lion. Nature (Lond) 242:530–513

Sumich JL (1983) Swimming velocities, breathing patterns, and estimated costs of locomotion in migrating gray whales, *Eschrictius robustus*. Can J Zool 61:647–652

Tenney SM, Remmers JE (1963) Comparative quantitative morphology of the mammalian lung: Diffusing area. Nature (Lond) 197:54–56

Tenney SM, Tenney JB (1970) Quantitative morphology of cold–blooded lungs: Amphibia and reptilia. Respir Physiol 9:197–215

Tenney SM, Bartlett D Jr, Farber JP, Remmers JE (1974) Mechanics of the respiratory cycle in the green turtle (*Chelonia mydas*). Respir Physiol 22:361–368

Teruoka G (1932) Die Ama und ihre Arbeit. Arbeitsphysiologie 5: 239–251

Thomas DP, Fregin GF (1981) Cardiorespiratory and metabolic responses to treadmill exercise in the horse. J Appl Physiol 50:864–868

Trillmich F, Kooyman GL, Majluf P, Sanchez-Grinan M (1986) Attendance and diving behavior of South American fur seals during El Nino in 1983. In: Gentry RL, Kooyman, GL (eds) Fur seals: Maternal strategies on land and at sea. Princeton University Press, Princeton, pp 153–167

Trillmich KGK, Trillmich F (1986) Foraging strategies of the marine iguana, *Amblyrhynchus cristatus*. Behav Ecol Sociobiol 18:259–266

Trivelpiece WZ, Bengtson JL, Trivelpiece SG, Volkman NJ (1986) Foraging behavior of Gentoo and Chinstrap penguins as determined by new radiotelemetry techniques. Auk 103:777–781

Valtin H (1973) Renal function: Mechanisms preserving fluid and solute balance in health. Little, Brown, Boston, 253 pp

Vatner SF, Higgins CB, Millard RW, Franklin D (1974) Role of the spleen in the peripheral vascular response to severe exercise in untethered dogs. Cardiovasc Res 8:276–282

Vennard JK, Street RL (1976) Elementary fluid mechanics. Wiley and Sons, New York, 740 pp

Viamonte M, Morgane PJ, Galiiano RE, Nagel EL, McFarland WL (1968) Angiography in the living dolphin and observations on blood supply to the brain. Am J Physiol 214:1225–1249

Videler J, Kamermans P (1985) Dolphin swimming performance: Differences between upstroke and downstroke. Aquat Mamm 11:46–52

Vik-Mo H, Mjos OD (1981) Influence of free fatty acids on myocardial oxygen consumption and ischemic injury. Am J Physiol 48:361–365

Vogel S (1981) Life in moving fluids: The physical biology of flow. Willard Grant, Boston, 352 pp

Wahren J (1977) Glucose turnover during exercise in man. Ann NY Acad Sci 301:45–55

Webb PW (1975) Hydrodynamics and energetics of fish propulsion. Bull Fish Res Board Canada 190:1–159

Webb PW (1984) Form and function in fish swimming. Sci Am 251: 72–82

Weber RE, Hemmingsen EA, Johansen K (1974) Functional and biochemical studies of penguin myoglobins. Comp Biochem Physiol 49:197–214

Weihs D (1973) Optimal fish cruising speed. Nature (Lond) 245:48–50

Weihs D (1974) Energetic advantages of burst swimming of fish. J Theor Biol 48:215–229

Weinheimer CJ, Pendergast DR, Spotila JR, Wilson DR, Standora EA (1982) Peripheral circulation in *Alligator mississipiensis*: Effects of diving, fear, movement, investigator activities, and temperature. J Comp Physiol 148:57–63

West JB (1974) Respiratory physiology: The essentials. Waverly, Baltimore, 185 pp

West JB (1977 a) Pulmonary pathophysiology. Williams and Wilkins, Baltimore, 227 pp

West JB (1977 b) Blood flow. In: Regional differences in the lung. Academic Press, New York, 85 pp

White JR, Harkness DR, Isaaks RE, Duffield DA (1976) Some studies on blood of the Florida manatee (*Trichechus manatus laterostris*). Comp Biochem Physiol 55:413–417

White FN, Ikeda M, Elsner RW (1973) Adrenergic innervation of large arteries in the seal. Comp Gen Physiol 4:271–276

Williams TM, Kooyman GL (1985) Swimming performance and hydrodynamic characteristics of the harbor seal (*Phoca vitulina*). Physiol Zool 58:576–589

Winter PM, Miller JN (1985) Anesthesiology. Sci Am 252:124–131

Woakes AJ, Butler PJ (1983) Swimming and diving in tufted ducks, *Aythya fuligula*, with particular reference to heart rate and gas exchange. J Exp Biol 107:311–329

Wood JD (1975) Oxygen toxicity. In: Bennett PB, Elliott DH (eds) The physiology and medicine of diving and compressed air work. Williams and Wilkins, Baltimore, pp 166–184

Wood SC, Gatz RN, Glass ML (1984) Oxygen transport in the green sea turtle. J Comp Physiol 154:275–280

Wood SC, Johansen K (1974) Respiratory adaptations to diving in the Nile Monitor lizard (*Varanus niloticus*). J Comp Physiol 89:145–158

Wood SC, Lenfant C (1976) Respiration: Mechanics, control and gas exchange. In: Gans C, Dawson WR (eds) Biology of the reptilia. Academic Press, New York, pp 225–274

Zagaeski M (1986) Some observations on the prey stunning hypothesis. Mar Mamm Sci 3:275–279

Zapol WM (1987) Diving adaptations of the Weddell seal. Sci Am 256:100–107

Zapol WM, Liggins GC, Schneider RC, Qvist J, Snider MT, Creasy RK, Hochachka PW (1979) Regional blood flow during simulated diving in the conscious Weddell seal. J Appl Physiol 47:968–973

Name Index

Ackerman, R.A.   95
Albers, C.   126
Aleksuik, M.   57
Aleyev, Y.U.G.   129, 132, 137
Andersen, H.T.   15, 16, 125
Angel-James, J.E.   78
Aschoff, S.   119, 120
Ashwell-Erickson, S.   122
Astrand, P.   90, 91, 122, 125, 127
Au, D.   139

Barcroft, J.   58, 59
Bartholomew, G.A.   119, 120
Baudinette, R.V.   148
Bauer, R.E.   135
Behnke, A.R.   29, 37
Belanger, L.F.   170
Belkin, D.A.   7–9, 11, 68, 82
Bello, M.A.   89, 91
Belloni, F.L.   93, 94
Bennett, A.F.   120
Bennett, P.B.   29, 30
Bentley, T.B.   146
Berkson, H.   17, 20, 50, 53, 95
Bert, P.   15
Bing, R.S.   85
Blake, R.W.   129, 132, 137, 138, 140
Blix, A.S.   18, 20, 21, 23, 169
Bohr, C.   110, 111
Booth, F.W.   93
Bradley, S.E.   21, 85
Brauer, R.W.   30, 31
Bron, K.M.   21
Brooks, G.A.   92, 104
Bryden, M.M.   24, 57, 62, 90, 122
Burggren, W.W.   49, 111
Butler, P.J.   1, 19, 67, 69–72, 81, 121, 147, 148

Campbell, W.B.   68, 75, 76
Campbell, S.D.   37
Carey, C.R.   34
Castellini, M.A.   59, 99, 107, 158
Clarke, M.R.   165
Cornford, N.E.   135
Costa, D.P.   124, 141
Costello, R.R.   122, 123
Cousteau, J.Y.   29

Craig, A.B.   33, 35, 39, 122
Croft, R.   39
Croll, D.A.   140
Cross, E.R.   47
Croxall, J.P.   156
Curray, J.   38

Davis, R.W.   65, 85, 86, 92, 99, 122, 127, 140,
    141, 147, 148
Dawson, W.R.   4, 120, 152
de Burgh, D.   78, 80
Djojosugito, A.M.   23
Dormer, K.J.   21
Dolphin, W.F.   122
Drent, R.H.   120
Dunlap, C.E.   49
Dykes, R.W.   78

Eckert, S.A.   38, 53, 153, 155
Eckert, K.L.   38, 153, 155
Eliassen, E.   97
Elsner, R.W.   1, 11–13, 17, 18, 20, 22, 25,
    62, 64, 67, 72–74, 84, 98, 122
Evans, W.E.   156

Falke, K.J.   42
Fedak, M.A.   81
Feldkamp, S.D.   123, 158, 165
Felger, R.S.   7
Folkow, B.   1, 18, 20, 23, 82–84, 90
Frair, W.   155
Fregin, G.F.   58
Frohlinger, A.   57
Fujise, Y.   59
Furilla, R.A.   15, 69, 70

Gabbott, G.R.J.   69
Gabrielsen, G.W.   69
Galliano, R.E.   27
Gallivan, G.J.   73
Gans, C.   68
Gatten, R.E.   8, 96
Gaunt, G.S.   68
Gentry, R.L.   1, 141, 158, 162
Gill, P.   148
Gleeson, T.T.   94
Gollnick, P.D.   90–92

193

Subject Index

mucous 135
murre 4, 38
  common 156
muscle 53, 57, 60–64, 67, 76, 81, 87, 119, 125, 126, 137
musk turtle 9
muskrats 57
Mustelidae 4
myocardium 127
myoglobin 60, 61, 64, 90
myosin 90

narcosis 29, 30, 37, 43, 44, 48, 52
nektonic 132
neonates 11, 59
nerve 69, 78
newts 30
Nile monitor lizard 111
nitrogen 29, 43
norepinephrine 93
northern elephant seal 2, 38, 44, 48, 53, 59, 62, 64, 103, 106, 149, 158, 159, 161, 162, 166

Odobenidae 4
olive Ridley sea turtle 4, 38, 41
Orcinus orca, see killer whale
Otariidae 4
oxygen 72, 82, 92, 99, 110, 115
  capacity 118
  consumption 46, 110, 113, 119, 125–127
  content 33, 35, 115–118
  dissociation curve 110–115
  toxicity 31, 37

packed cell volume (PCV) 58, 59, 60, 117
PACO2 111, 114
painted turtle 8, 67
pancreas 86
PaO2 81, 110–117
PAO2 110–117
Pasteur effect, reversed 126
PCO2 93, 107, 109–114, 126
pearl diver 47
pectoral fin 138
pelagic 5, 48, 60
Pelamis platurus, see yellow-bellied sea snake
perfusion 18, 47
peripheral vascular resistance 19
Phalacrocorax auritus, see cormorant
Phalocrocoracidae 4
Phoca vitulina, see harbor seal
Phocidae 4, 56
phocids 38, 56, 60
phrenic nerve 25
Physeteridae 169
pigeon 56
pike 135

pilot whale 137, 165
Pinnipedia 60
plasma 87, 96, 99
  volume 57, 58
plastron 41
plesiosaurs 4
pochard 71
polysaccharide 85
post-caval sphincter 11
posterior vena cava 24–26
pregnant 22
pressure 18, 21, 22, 68, 78, 81
primates 30
propulsion 135
Pseudemys scripta, concinna 67, 78
  scripta (P. scripta), see freshwater turtle
Pseudorca crassidens 24
psychogenic 69, 81
pulmonary 67, 78
PvO2 112, 115
pyruvate 10, 11, 92

radioisotopes 25
radionuclides 18
rat 84, 90–93, 97–102, 105, 106, 108, 119–128
razorbills 97
rectal temperature 125, 126
red blood cells 57, 60
redhead diving duck 69
reflex 67, 69, 78, 80, 82
  response 15, 16
renal 118
  artery 21
  blood flow 83–85, 87
reptiles 30, 32, 38, 41, 48–51
resting metabolism 143, 146, 148, 153
  metabolic rate (RMR) 119–125, 128
rete mirabile 27, 170
Reynolds number 131
rhesus monkey 12

scuba 29, 118
sea otter 4, 54, 114
  lion 129, 136, 137
seizures 32
shags 97
sheep 37
shunt 23, 31, 43, 44, 48, 49, 51, 52, 81, 82, 109–112, 153
sinus 98
Sirenia 4
Sirenidae 54
skeletal muscle 63, 89, 90, 92–94, 106–108
skin 67, 76, 109, 131, 135
snake 38, 41, 48, 52
sooty tern 148
southern elephant seal 57, 59, 62, 122